Introduction to carbohydrate chemistry

GUTHRIE AND HONEYMAN'S

Introduction to carbohydrate chemistry

BY
R. D. GUTHRIE
FOUNDATION PROFESSOR OF CHEMISTRY
GRIFFITH UNIVERSITY, BRISBANE

FOURTH EDITION

CLARENDON PRESS · OXFORD
1974

Oxford University Press, Ely House, London W.1

GLASGOW   NEW YORK   TORONTO   MELBOURNE   WELLINGTON
CAPE TOWN   IBADAN   NAIROBI   DAR ES SALAAM   LUSAKA   ADDIS ABABA
DELHI   BOMBAY   CALCUTTA   MADRAS   KARACHI   LAHORE   DACCA
KUALA LUMPUR   SINGAPORE   HONG KONG   TOKYO

CASEBOUND ISBN 0 19 8551428
PAPERBACK ISBN 0 19 8551436

© OXFORD UNIVERSITY PRESS 1968, 1974

FIRST EDITION 1948
SECOND EDITION 1964
THIRD EDITION 1968
FOURTH EDITION 1974

QD
321
.28x
1974

PRINTED IN GREAT BRITAIN
BY J. W. ARROWSMITH LTD
BRISTOL BS3 2NT

Preface to the fourth edition

MY aim in completely re-writing this text has remained that of the previous editions: to provide an up-to-date introductory text of a size attractive to the prospective student purchaser. It is hoped too that it will be of use to the established chemist or biochemist, seeking to move into the general area of carbohydrate chemistry.

Because of developments that have occurred in the past few years and of the wish to keep the book to the length of previous editions, some classical carbohydrate chemistry has been deleted, and other areas of the subject dealt with briefly. I have tried to concentrate on areas of current interest and those that are likely to develop in the future.

The re-writing and re-structuring of the book has allowed glycosides to be treated more fully, and a brief introduction to the chemotherapeutically important carbohydrate antibiotics to be included. The physical methods chapter includes a section on ^{13}C nuclear magnetic resonance spectroscopy, a technique that will become important during the next few years. The final chapter is an effort to let the reader see the application of the chemistry described in the book to real problems.

A major departure from previous editions and from other similar carbohydrate texts is the use of Mills'-type formulae. Use of this representation for carbohydrate structures brings this area of chemistry into line with organic chemistry generally, and should help to emphasise that carbohydrate chemistry is not separate in any way from the remainder of that general field. Established carbohydrate chemists can be assured that the change in usage from Haworth formulae to Mills ones can be accomplished quickly and easily.

The whole text was read by Drs. J. Honeyman, N. A. Hughes, and I. Jenkins, and Chapter 14 by Dr. D. A. Rees. I thank them all for their helpful criticism and comments.

Most of the work on this book was done whilst the author was a member of the School of Molecular Sciences, University of Sussex, Brighton, U.K.

April, 1973   R.D.G.

72302

Preface to the third edition

OUR aims have remained the same as in previous editions. Most chapters have been slightly modified, and that on physical methods has been lengthened. We have also added compounds of current interest to the chapter on miscellaneous derivatives. We have endeavoured to keep the book the same length as the previous edition.

April 1968   R.D.G.
                                                                J.H.

Preface to the second edition

WE have tried to make the new edition concise and up-to-date, concentrating on important principles of carbohydrate chemistry. The increased use of physical methods and the interest in the shapes of molecules have caused us to add new chapters on these topics. No detailed references are given to the original literature but important review articles are listed at the end of appropriate chapters.

                                                                R.D.G.
                                                                J.H.

Preface to the first edition

THE aim of the author in writing this book has been to provide, in a compact form, an up-to-date account of the chemistry of a selected number of carbohydrates. No previous knowledge of the field is assumed, but the reader requires to be familiar with the elements of aliphatic and aromatic chemistry. The monosaccharides are considered in detail in order to provide an adequate basis for studying farther the more complex carbohydrates. In addition to the crystalline di-, tri-, and tetra-saccharides, a few of the simpler polysaccharides of colloidal dimensions are discussed. The account of these does not claim to be exhaustive, but it is hoped that the introduction given here will enable the student to read, with greater ease, some of the more complete specialized studies now available.

While the material included provides a suitable course for students reading for an Honours degree in Chemistry, it is hoped that the book will appeal to a wider public. Academic and industrial chemists in related branches of organic chemistry who wish to keep abreast of recent developments in theory and technique should find the book of value.

It has been considered inappropriate to include numerous references to the original literature, but, in the case of many of the more important discoveries, the name of the author and the date are included. In this way the reader is introduced to most of the chief workers in the carbohydrate field. The original paper is easily located, when desired, by referring to the author index of *British (Chemical) Abstracts* or *Chemical Abstracts* for the appropriate year.

The author has to thank many friends for their assistance in making it possible for him to write this book. For help in the early, most difficult, years of his career he is grateful to Dr. James Scott, formerly Science Master at Riverside School, Stirling, and to Sir James C. Irvine, Principal and Vice-Chancellor of the University of St. Andrews. Professor D. H. Hey of the University of London, King's College, read the manuscript and made many valuable suggestions. The author has had the indispensable aid and co-operation of his wife in the preparation of this work for the press.

Chiswick, 1947

Contents

1 General introduction

CARBOHYDRATES are among the most abundant organic compounds found in natural sources; they are widespread in both plants and animals. They range from small molecules with molecular weights a little over one hundred to high polymers with molecular weights well above one million. Carbohydrates appear at an early stage in the conversion of carbon dioxide into organic compounds by plants, which build up carbohydrates from carbon dioxide and water by photosynthesis. Animals have no way of synthesizing carbohydrates from carbon dioxide and rely on plants for their supply. The carbohydrates are then converted into other organic materials by a variety of biosynthetic pathways.

Carbohydrates serve as sources of energy (sugars) and stores of energy (starch and glycogen); they also form a major portion of the supporting tissue of plants (cellulose) and of some animals (crabs, lobsters, etc.); they play a basic role as part of the nucleic acids DNA and RNA. Others are found as components of a variety of natural products, such as bacterial cell walls, blood group substances, and cartilage. In recent years a new growth point has developed, in that there are many carbohydrate substances, most of which contain amino-sugars, that are antibiotics.

The name 'carbohydrate' arose from the mistaken belief that substances of this kind were hydrates of carbon, since the molecular formula of many could be expressed in the form $C_x(H_2O)_y$, for example glucose ($C_6H_{12}O_6$), sucrose ($C_{12}H_{22}O_{11}$).

Carbohydrates can be divided into two broad groups: sugars and polysaccharides.

Sugars are sweet, crystalline, soluble in water. Their molecular weight is known exactly and is invariable for a given substance. The simplest sugars are the *monosaccharides*, which chemically are polyhydroxyaldehydes or polyhydroxyketones. A *disaccharide* is built up from two monosaccharide units with the formal elimination of one molecule of water; a *trisaccharide* from three units less two molecules of water, etc. The monosaccharides thus constitute the main building blocks of carbohydrates. A disaccharide, trisaccharide, etc., can be hydrolysed by dilute aqueous acid to the constituent monosaccharides.

Oligosaccharide is the group name given to sugars containing from two to ten monosaccharide units.

Compounds closely related to sugars and generally studied alongside them though not strictly members of that class, are the polyhydroxyalkanes (alditols) and polyhydroxycycloalkanes (cyclitols). *Polysaccharides* (glycans) are polymeric substances. The molecular weight of the molecules in a given sample will not all necessarily be of the same size. They are built from thousands of monosaccharide units, in a way similar to the simpler di- and tri-saccharides. Complete hydrolysis of a polysaccharide with dilute aqueous acid breaks it down to its constituent monosaccharides.

Nomenclature

A brief introduction to the nomenclature of carbohydrates is necessary before proceeding further. As stated above, the basic sugar units are the monosaccharides which are either polyhydroxyaldehydes called *aldoses*, or polyhydroxyketones called *ketoses*. Generally an indication of the number of carbon atoms present is given so that glucose ($C_6H_{12}O_6$) which is an aldose is more fully described as an *aldohexose*: fructose (also $C_6H_{12}O_6$) is a *ketohexose*. Replacement of a hydroxyl group by hydrogen, i.e., the nominal change $\geqslant$C—OH to $\geqslant$C—H is shown by the prefix *deoxy*. Thus rhamnose ($C_6H_{12}O_5$) is a deoxyaldohexose. Replacement of a hydroxyl group by any other substituent is formally regarded as going *via* the deoxy sugar. Thus a derivative in which the change $\geqslant$C—OH to $\geqslant$C—NH$_2$ has been made at one hydroxyl group, is called an amino-deoxy-sugar. Formation of ether groups, most commonly methyl, ($\geqslant$C—OH to $\geqslant$COMe) is shown by adding '*O*-methyl-' to the front of the name preceded by the number of the carbon atom whose hydroxyl group has been etherified. Esters, for example, acetates ($\geqslant$C—OH to $\geqslant$COCOCH$_3$), are shown by adding either '*O*-acetyl-' before the name or 'acetate' after it, in each case again preceding it with the appropriate carbon number. The nomenclature of other derivatives will be developed as required.

2 Structure and configuration of the monosaccharides

Structure of the monosaccharides

The commonest monosaccharide is glucose (sometimes trivially called dextrose). It occurs free in the juice of fruits and in honey, and may also be obtained by acid hydrolysis of many other carbohydrates, such as cellulose and starch. The establishment of glucose as a polyhydroxyaldehyde ($C_6H_{12}O_6$, an aldohexose), followed from chemical evidence; in particular its reaction with hydrogen cyanide, hydrolysis of the cyanohydrin to the carboxylic acid, and its reduction with phosphorus and hydriodic acid to give heptanoic acid (Scheme 2.1).

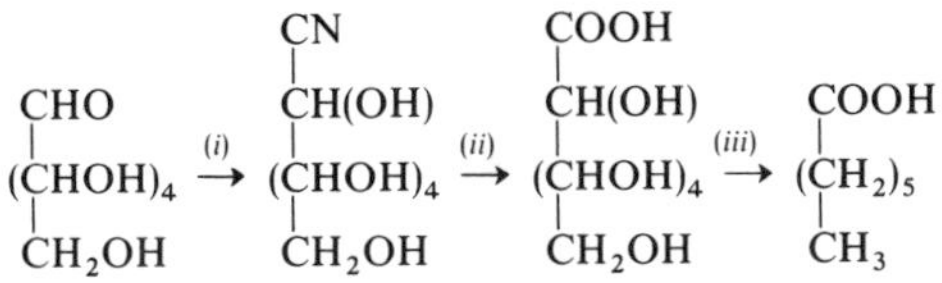

$$
\begin{array}{ccccc}
 & \text{CN} & \text{COOH} & \\
 & | & | & \\
\text{CHO} & \text{CH(OH)} & \text{CH(OH)} & \text{COOH} \\
| & \overset{(i)}{\rightarrow} \quad | & \overset{(ii)}{\rightarrow} \quad | & \overset{(iii)}{\rightarrow} \quad | \\
(\text{CHOH})_4 & (\text{CHOH})_4 & (\text{CHOH})_4 & (\text{CH}_2)_5 \\
| & | & | & | \\
\text{CH}_2\text{OH} & \text{CH}_2\text{OH} & \text{CH}_2\text{OH} & \text{CH}_3 \\
\end{array}
$$

2.1

Reagents: (i) HCN; (ii) HO^-; (iii) HI–P

SCHEME 2.1

Glucose is therefore a 2,3,4,5,6-pentahydroxyhexanal **2.1**. Other evidence in keeping with structure **2.1** was the formation of aldehyde derivatives, reduction to a hexahydroxyhexane or alditol, called glucitol **2.2**, and oxidation to a pentahydroxy acid, gluconic acid **2.3**.

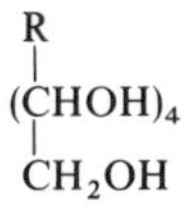

$$
\begin{array}{c}
\text{R} \\
| \\
(\text{CHOH})_4 \\
| \\
\text{CH}_2\text{OH} \\
\end{array}
$$

2.2 R = CH_2OH
2.3 R = COOH

Another common monosaccharide is fructose (or laevulose) ($C_6H_{12}O_6$) often found in association with glucose. Application of the sequence of reactions that was used in Scheme 2.1 to fructose gave 2-methyl-hexanoic acid (Scheme 2.2). Fructose must therefore be a polyhydroxyketone, a ketohexose, and must have gross structure **2.4**.

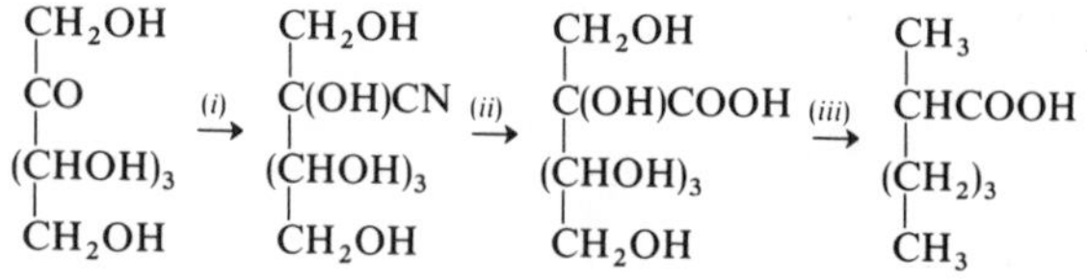

$$
\begin{array}{cccc}
\text{CH}_2\text{OH} & \text{CH}_2\text{OH} & \text{CH}_2\text{OH} & \text{CH}_3 \\
| & | & | & | \\
\text{CO} & \text{C(OH)CN} & \text{C(OH)COOH} & \text{CHCOOH} \\
| & | & | & | \\
(\text{CHOH})_3 & (\text{CHOH})_3 & (\text{CHOH})_3 & (\text{CH}_2)_3 \\
| & | & | & | \\
\text{CH}_2\text{OH} & \text{CH}_2\text{OH} & \text{CH}_2\text{OH} & \text{CH}_3
\end{array}
$$

with steps (i), (ii), (iii) between the structures.

2.4   Reagents: as Scheme 2.1

SCHEME 2.2

The structures of all monosaccharides may be established similarly, and they are summarized in general terms in Table 2.1. Note that the numbering of the carbon chain follows the usual rules of organic nomenclature such that the aldehyde carbon is numbered 1 and is referred to as C-1, and that the carbonyl group in ketoses is at C-2

TABLE 2.1

Monosaccharide structures

	Carbon atom Number	Trioses	Tetroses	Pentoses	Hexoses
Aldoses	1	CHO	CHO	CHO	CHO
	2	*CHOH	*CHOH	*CHOH	*CHOH
	3	CH₂OH	*CHOH	*CHOH	*CHOH
	4		CH₂OH	*CHOH	*CHOH
	5			CH₂OH	*CHOH
	6				CH₂OH
Ketoses	1	CH₂OH	CH₂OH	CH₂OH	CH₂OH
	2	CO	CO	CO	CO
	3	CH₂OH	*CHOH	*CHOH	*CHOH
	4		CH₂OH	*CHOH	*CHOH
	5			CH₂OH	*CHOH
	6				CH₂OH

*asymmetric carbon atom

Configuration of the monosaccharides

The solution to the problem of the configuration of the mono-saccharides is one of the classics of organic chemistry. This achievement may seem slight in modern terms, but considering the tools available, it was a masterpiece.

Emil Fischer, during the period 1891 to 1896, tackled this problem using the then new ideas on stereochemistry of Le Bel and van't Hoff. Stereoisomers which differ only in the arrangement of groups around a carbon atom, are also optical isomers in that the two stereoisomers rotate the plane of plane-polarized light an equal amount, but in opposite directions. Such isomers contain what is called an 'asymmetric carbon atom', or a *chiral centre*. In glyceraldehyde **2.5**, this chiral centre is C-2, marked with an asterisk. Fischer considered the groups around a chiral centre to be at the vertices of a tetrahedron as shown in **2.6** and **2.7** for the two possible stereoisomers of glyceraldehyde **2.5**.

For any system with n chiral carbon atoms, there are 2^n stereoisomers, composed of 2^{n-1} enantiomeric (mirror-image) pairs. Table 2.2 shows the total number of stereoisomers possible for the trioses to the hexoses. Note that ketoses have one less chiral centre than aldoses with the same number of carbon atoms.

TABLE 2.2

Sugar	Number of chiral centres	Number of optical isomers	Number of enantiomorphic pairs
Aldotrioses	1	2	1
Aldotetroses and ketopentoses	2	4	2
Aldopentoses and ketohexoses	3	8	4
Aldohexoses	4	16	8

Fischer devised planar representations of the monosaccharides by considering the lower edges of the tetrahedra forming the carbon chain to be in a straight line as shown in **2.8**. (The dotted line shows the position of the lower edge.) Projection of this form on to the plane below it gives **2.9**—a 'Fischer projection formula'.

$$
\begin{array}{cc}
\text{CHO} & \text{CHO} \\
\text{H}\longleftrightarrow\text{OH} & \text{H}\!\!-\!\!\text{OH} \\
\text{H}\longleftrightarrow\text{OH} & \text{H}\!\!-\!\!\text{OH} \\
\text{HO}\longleftrightarrow\text{H} & \text{HO}\!\!-\!\!\text{H} \\
\text{CH}_2\text{OH} & \text{CH}_2\text{OH} \\
\textbf{2.8} & \textbf{2.9}
\end{array}
$$

The disadvantages of Fischer formulae cannot be overstressed. In modern terms, such projections are those of the least stable, fully eclipsed conformation (see p. 18). The apparent *cis* or *trans* relationship of vicinal groups in the projection does not necessarily occur in the actual shape taken up by the molecule. Nonetheless, Fischer projections are the only way to represent in two dimensions an acyclic molecule with several chiral centres. Fischer differentiated between the members of an enantiomeric pair of monosaccharides by the use of *d* (read as 'dextro') and *l* (read as 'laevo') depending on the nature of their optical rotation. Rosanoff (1906) appreciated that some *configurational* feature, and not rotation, must be the one used for differentiation and took as reference the hydroxyl group on the penultimate carbon atom in the chain (that is, the highest numbered chiral centre). If this was to the right in the Fischer formula he called it a δ-sugar, if to the left, a λ-sugar. These symbols were later changed to D (read as 'dee') and L (read as 'ell'), and are now universally used. It must be emphasized that the symbols D and L are *completely independent of optical rotation* and show only the chirality of the molecule. If necessary, the sign of rotation is shown by (+) or (−), placed between the D (or L) and the sugar name.

Structure **2.10** was selected arbitrarily by Rosanoff for (+)-glyceraldehyde (from the two possibilities) and named D, and then this *convention* was used as a basis for the other monosaccharides. It had to be a convention because the *absolute* configuration of glyceraldehyde was not known to him. Thus **2.10** is D-(+)-glyceraldehyde and **2.11**, L-(−)-glyceraldehyde. The chirality of other molecules was then established by relating them to D- or L-glyceraldehyde. Thus L-(+)-arabinose has the structure **2.12**. It was not until 1951 that Bijovet showed by X-ray structural studies that the absolute configuration of D-glyceraldehyde was identical to the structure arbitrarily assigned to it by Rosanoff. Thus immediately all relative configurations became absolute configurations.

CHO	CHO	CHO	CHO
H—OH	HO—H	H—OH	H—OH
CH$_2$OH	CH$_2$OH	HO—H	HO—H
		HO—H	H—OH
		CH$_2$OH	H—OH
			CH$_2$OH
2.10	**2.11**	**2.12**	**2.13**

Fischer's elegant solution to the problem of the configuration of D-glucose **2.13** is detailed in Appendix A. Similar methods were used for the configurations of the other monosaccharides. The configurations of the aldotrioses to aldohexoses in the D-series are shown inside the back cover.

When Fischer began his work with carbohydrates (1886) the only monosaccharides known were L-arabinose, D-glucose, D-galactose, D-fructose and L-sorbose, all isolated from natural sources. When Fischer died in 1919, fourteen of the possible sixteen hexoses had been synthesized, the remaining two followed in 1934 (Austin). Some common natural monosaccharides are listed in Table 2.3.

TABLE 2.3

Physical properties of some naturally occurring monosaccharides

	Main source	m.p. °C	$[\alpha]_D$ in water degrees
D-Ribose	Nucleic acids	87	$-23 \rightarrow -24$
D-Xylose	Xylan	145	$+94 \rightarrow +19$
L-Arabinose	Gums	160	$+190 \rightarrow +105$
D-Glucose	Starch	146	$+112 \rightarrow +53$
D-Mannose	Mannan	133	$+29 \rightarrow +14$
D-Galactose	Gums	167	$+150 \rightarrow +80$
L-Rhamnose	Glycosides	124	$-9 \rightarrow -8$
L-Fucose	Seaweeds	145	$-153 \rightarrow -76$
D-Fructose	Insulin	102	$-132 \rightarrow -94$
L-Sorbose	Rowans	159	$-44 \rightarrow -43$
D-Tagatose	Gums	134	$-3 \rightarrow -5$

Total synthesis of monosaccharides

This possibility has intrigued organic chemists for nearly a century. The monosaccharides could be made nominally from formaldehyde (HCHO or CH_2O) [cf. $C_x(H_2O)_y$ for sugars] and the first attempts by Butleroff (1861) and Leon (1886) using the action of dilute alkali on formaldehyde gave a sweet syrup called *formose*; this is a mixture of aldoses and ketoses. Another synthetic sugar

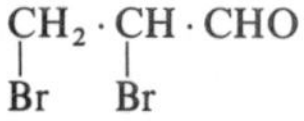

$$\underset{\substack{| \\ Br}}{CH_2} \cdot \underset{\substack{| \\ Br}}{CH} \cdot CHO$$

2.14

acrose was prepared by Fischer from the action of alkali on acrolein dibromide (2,3-dibromoproprionaldehyde) **2.14**. DL-Glyceraldehyde is the first product which is then isomerized partially to dihydroxyacetone. Aldol condensation between these two C_3-units yields a mixture of ketohexoses. DL-Fructose and DL-sorbose have been isolated from acrose as phenylosazones (Schmitz, 1913). Recent research in this area has tried to mimic ways in which formaldehyde may have been converted to sugars under prebiotic conditions (that is, conditions of primitive Earth).

Ring-structures of the monosaccharides

Any aldehyde or ketone can form a hemiacetal with an alcohol. If within one molecule there is a hydroxyl group and a carbonyl group in suitable stereochemical relationship, then there is an equilibrium between the open-chain and the cyclic or lactol form in solution (Scheme 2.3). The position of this equilibrium obviously depends on the thermodynamic stability of the cyclic versus the acyclic form.

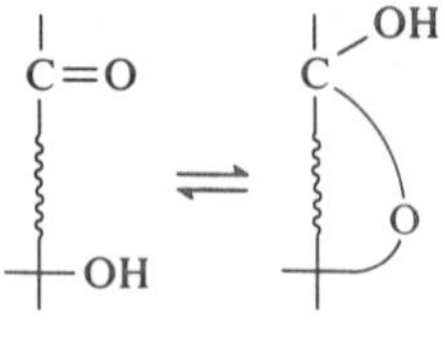

SCHEME 2.3

For a monosaccharide, for example, D-glucose **2.13** there is obviously the possibility for ring-formation to occur, and in fact it does. Inspection of **2.13** shows that three-, four-, five-, six-, or seven-membered lactol rings are in theory all possible.

Many chemical properties of D-glucose suggest that it is not a normal aldehyde. For example, it does not react with Schiff's reagent; its acetylated and benzoylated derivatives do not react with the usual carbonyl reagents such as hydroxylamine. Spectral evidence also supports a ring structure by the absence of a carbonyl band in the ultraviolet or infrared spectra.

If a ring is present, can all the various possibilities mentioned above occur? As expected from general organic chemistry, it is found that the lactol ring in monosaccharides is five- or six-membered, rarely seven-membered, and never three- or four-membered.

In a chiral molecule such as D-glucose the formation of the lactol ring creates a new chiral centre at C-1 **2.15**. (In a ketose such as D-fructose, this new centre is formed at C-2 **2.16**.) Therefore, for any particular size of lactol ring there are two C-1 epimers, called *anomers*. They are distinguished by the use of the prefixes α and β. The former is used when the hydroxyl group at the anomeric centre (that is at C-1 in aldoses and at C-2 in ketoses) is on the same side of the Fischer projection formula as the highest numbered chiral centre (i.e., C-5 in glucose) and β is used when the anomeric hydroxyl group is on the opposite side. Thus **2.15** is an α-D-glucose, and **2.16** is a β-D-fructose.

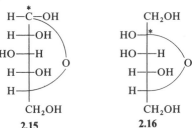

2.15   2.16

Haworth coined the use of the terms *furanose* and *pyranose* for five- and six-membered rings, respectively. These are based on the names for the corresponding parent heterocyclic compounds furan **2.17** and pyran **2.18**. These names should be incorporated into the description of a sugar or a derivative of a sugar if the ring size is

2.17   2.18

known. Thus the common crystalline form of glucose is α-D-gluco-pyranose **2.15**. Seven-membered rings are indicated by the use of the term *septanose*.

Haworth also devised representations for the pyranose and furanose sugars. These *Haworth formulae* are perspective diagrams for the ring, which is drawn as in the plane at right angles to that of the paper, with the attached atoms or groups lying parallel to the plane of the paper. Thus **2.19** and **2.20** are two representations of α-D-glucopyranose; the former is the commonly used form, the convention being that the ring-oxygen is at the back on the right. *β*-D-Glucopyranose is represented by **2.21**. L-Sugars are the mirror-images of the D-forms. Thus *β*-L-glucopyranose can be represented as **2.22** or **2.23**; the latter is the generally used form. Haworth formulae can be used for furanose sugars as well. For example, *β*-D-ribofuranose is **2.24**, *β*-D-fructofuranose is **2.25**.

2.19   **2.20**   **2.21**

α-D-Glucopyranose   *β*-D-Glucopyranose

2.22   **2.23**

2.24   **2.25**

β-D-Ribofuranose   *β*-D-Fructofuranose

An alternative representation of pyranose and furanose sugars is that proposed by Mills, and is in keeping with the formulae used by organic chemists generally. In this system heavy lines denote groups above the plane of the ring (the β-face) and dotted lines denote groups below that plane (the α-face). Groups of uncertain or unknown orientation are shown by a wiggly line. Thus α-D-glucopyranose would be **2.26** (compare **2.19**), β-D-glucopyranose is **2.27** (compare **2.21**) and β-D-ribofuranose is **2.28** (compare **2.24**).

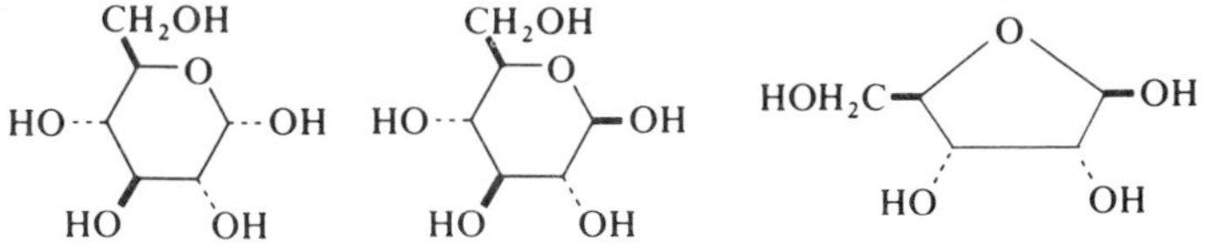

2.26	**2.27**	**2.28**
α-D-Glucopyranose	β-D-Glucopyranose	β-D-Ribofuranose

Mills-type formulae will in general be used in the remainder of this book. It is felt that they are easier to draw, are more meaningful than Haworth formulae, and bring carbohydrate chemistry into line with other natural product areas such as terpenes and steroids. It is hoped that their use in this book will encourage their more widespread use.

Solution equilibria: mutarotation

The optical rotation of a freshly prepared aqueous solution of α-D-glucopyranose changes rapidly until it reaches a constant value. This phenomenon is known as *mutarotation* and arises from the complex equilibria set up on dissolution of monosaccharides. These equilibria are shown in Scheme 2.4.

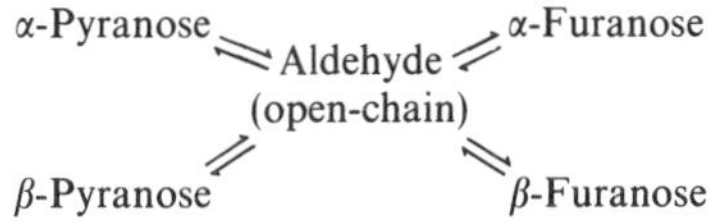

SCHEME 2.4

Thus α-D-glucopyranose, initial $[\alpha]_D + 113°$ (water) shows eventually a constant rotation of $+52.5°$. β-D-Glucopyranose shows an initial rotation of $+19°$ (water) rising to the same value of $+52.5°$.

The proportion of the five possible forms at equilibrium will vary widely from sugar to sugar depending on the thermodynamic stabilities of each. Generally the acyclic (aldehyde) form is only ever present to a very small extent and of the ring forms, the pyranoses are preponderant. For D-glucose in aqueous solution, the equilibrium mixture is essentially a mixture of the α- and β-pyranose forms. For other sugars, such as D-ribose, the furanoses form a significant proportion. Nuclear magnetic resonance methods are particularly useful in investigating these equilibria, as the four anomeric protons (α- and β-pyranose, α- and β-furanose) are usually assignable and their relative peak areas can be measured. A different equilibrium mixture is obtained in other solvents. Aprotic solvents, such as dimethyl sulphoxide, solvate hydroxyl groups less well than water and the percentage of furanose forms increases accordingly. Methylation of hydroxyl groups can have a similar effect.

Mutarotation generally shows general acid-base catalysis as shown in Scheme 2.5. For some sugars with essentially only a two components system (e.g. D-glucose) the mutarotation follows first order kinetics (Lowry, Hudson); for sugars with complex equilibria (e.g. D-fructose) simple first-order kinetics are not followed.

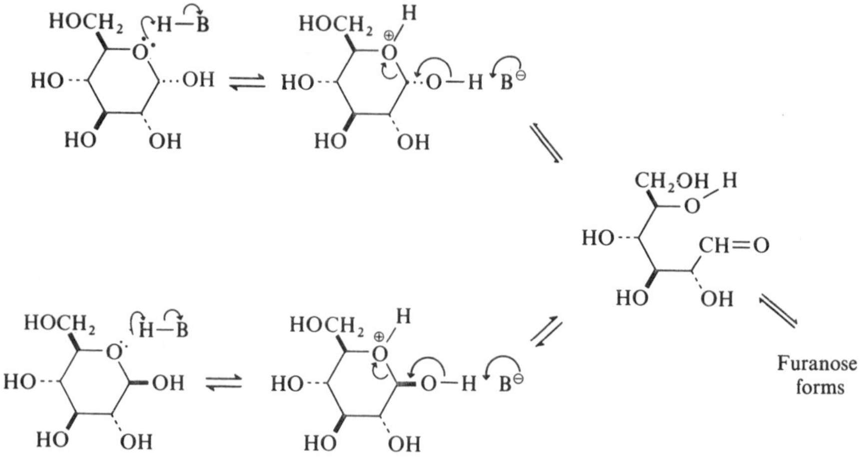

SCHEME 2.5

Although the concentrations of a particular form may be low, derivatives of it may be produced because of disturbance of the equilibrium. Hence, D-glucose with a reagent such as ethanethiol leads to a rapid and complete production of the acyclic diethyl dithioacetal (see p. 29).

3 Conformations of monosaccharides

THE word 'conformation' was introduced into chemistry by Haworth in 1929, though the idea of a non-flat, puckered ring for glucose had been utilized previously by Sponsler and Dore in their work on cellulose (1926). Detailed studies of the conformations of molecules have developed extensively since about 1950, particularly following the pioneer work of D. H. R. Barton (Nobel prize 1969).

The *conformations* of a molecule are the arrangements in space of the atoms of a single chemical structure (configuration); the arrangements are produced only by rotation about one or more single bonds and are not superimposable. The various conformations of a molecule are called *conformers*.

Full understanding of this topic is impossible without study of molecular models; to follow further points in this chapter Catalin, Dreiding, or similar atomic models are recommended.

Pyranose systems

Investigation of the cyclohexane molecule, has revealed that of the possible strainless forms the chair conformation **3.1** rather than a boat **3.2** is preferred.

3.1　　　　**3.2**

In the chair and boat conformers the bonds bearing substituents are either axial (*ax*), shown by full bonds in **3.1** and **3.2**, or equatorial (*eq*), shown by dotted bonds. In the conversion of one chair form into another the equatorial substituents become axial and *vice versa*.

In a solution of a simple monosubstituted cyclohexane derivative, the two possible chair conformers **3.3** and **3.4** are in equilibrium, but the preponderant form is **3.3** in which the substituent is equatorial with minimum non-bonded interactions with neighbouring hydrogen atoms. In general, the preferred conformer of a molecule is that with the greater number of bulky groups in equatorial positions.

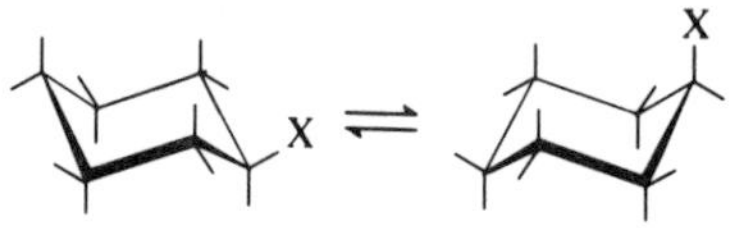

3.3   3.4

However, hydrogen-bonding or dipolar interactions can cause a molecule to exist preferentially in a conformation with axial substituents. For example, 5-hydroxy-1,3-dioxan exists predominantly in the conformation with the hydroxyl group axial **3.5**, rather than with the substituent equatorial **3.6** (Foster 1959).

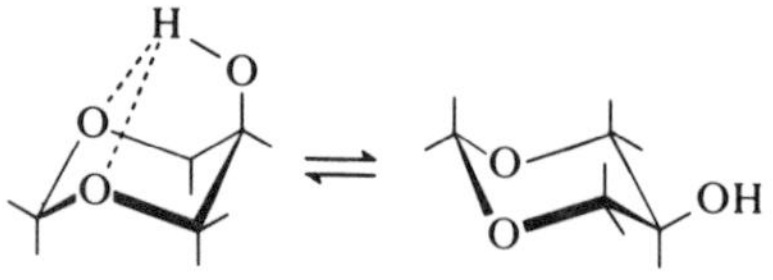

3.5   3.6

The replacement of one carbon atom of cyclohexane with an oxygen atom to give a pyranose ring does not cause any appreciable distortion, and in general the conformational principles described for cyclohexane are applicable. Pyranose rings carry many substituent groups between which non-bonded interactions are possible. Crystalline β-D-glucopyranose is **3.7**, where all the large groups are equatorial, rather than **3.8**, where they are axial. Other sugars have equatorial and axial substituents in each conformer; **3.9** shows the preferred form of α-D-mannopyranose.

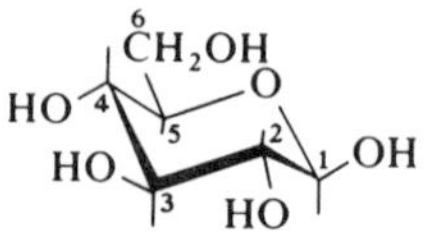

3.7   3.8   3.9

β-D-Glucopyranose   β-D-Glucopyranose   α-D-Mannopyranose

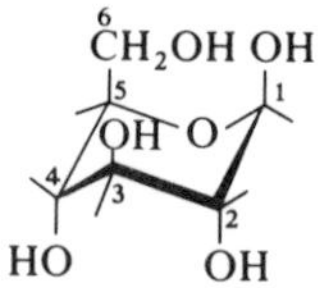

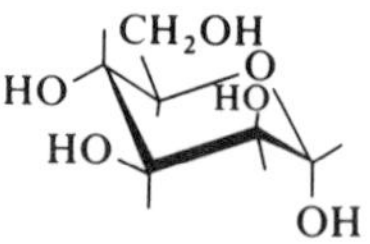

A *boat* conformation is flexible and is only one of six on a cycle of an infinite number of strainless forms. Most of these are not readily depictable on paper. A *twist* (or skew boat) form occurs halfway between each pair of boat forms. (Use of molecular models illustrates this point very clearly.) As shown in **3.10**, the six ring atoms are in two groups of three, each in one plane as shown, the central atom of each group being in the plane of the other group. An attempt to draw **3.10** in perspective is shown in **3.11**.

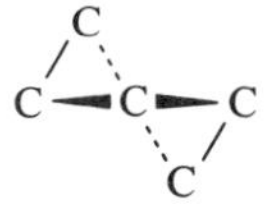

3.10 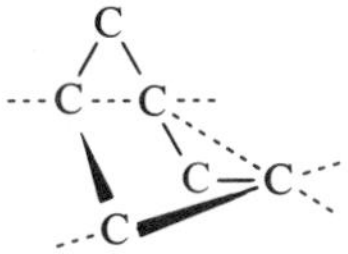   **3.11**

The *half-chair* conformation has to be considered when a six-membered ring contains either a double bond **3.12**, or an oxiran ring **3.13**, causing four adjacent atoms to be in one plane.

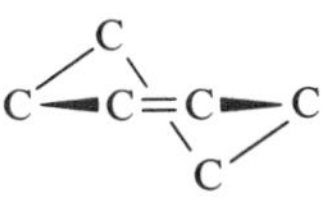

3.12 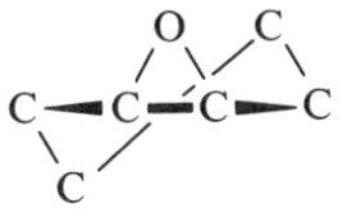   **3.13**

When considering part of a molecule of a conformation for stereochemical purposes, the dihedral angle (torsion angle) between groups on adjacent ring atoms is important. This is the angle between groups when viewed along the bond joining them, as in **3.14**. A circle is drawn in the projection diagram to represent the bond joining the two rings atoms (Newman projection).

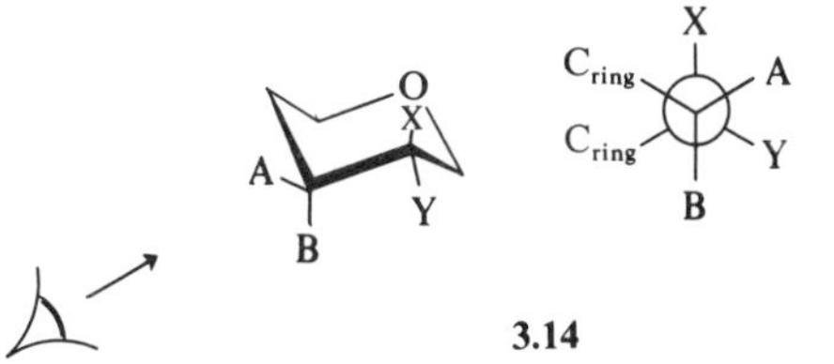

3.14

Nomenclature

There is no internationally agreed nomenclature system for pyranose conformers, though there is hope that this situation will change soon. The original system, due to Reeves (1946), is inadequate for some purposes, and other systems have been proposed (Guthrie 1958, Isbell and Tipson 1959). Reeves defined the shapes of chair forms of the pyranose ring, without reference to substituents, as shown in **3.15** and **3.16**. When the ring shape **3.16** is turned over through 180° it gives **3.17**, which is the mirror image of **3.15**. Therefore, the D- and L- forms of a sugar in the same conformation (that is with the same relative axial-equatorial arrangement of groups) have different symbols; C1 and 1C must, therefore, always be accompanied by D- or L-, otherwise they are without meaning.

Cl $^4C_1$ 1C $^1C_4$ 1C $^1C_4$

3.15   **3.16**   **3.17**

An alternative system which can be applied to chair forms as well as to other conformations, and which will probably become generally accepted, uses the initial letter of the shape, that is C for chair, S for skew boat etc., together with two numbers which show the out-of-plane atoms [see **3.15**, **3.16**, **3.17**]. The viewing face of the molecule is that from which the ring-numbering appears to go in a clockwise sequence. Thus β-D-glucopyranose in the $^4C_1$ or C1 conformation is shown in **3.7**, and in the $^1C_4$ or 1C conformation **3.8**.

Determination of conformation

The most powerful tool is proton magnetic resonance spectroscopy (see p. 98); the coupling constants between vicinal protons yield information about their relative stereochemistry.

In pre-n.m.r. days the pioneering work was carried out by Reeves (1946 on), who studied the optical rotation of copper complexes of monosaccharide derivatives, generally glycosides. He showed that vicinal diols would only act as ligands if the dihedral angle was between 0° and 60°. The former angle occurs only in boat conformations, the latter for *eq,eq* and *ax,eq* diols; *ax,ax* diols, having a dihedral angle of 180°, do not form complexes. Using this method Reeves was able to show that chairs were preferred to boats, and to decide which of the two chair forms was preferred. The method has recently been extended for use with vicinal amino-alcohols, and further by measuring the circular dichroism spectra of the copper complexes (Guthrie 1968) rather than the single wavelength method of Reeves.

Anomeric effect

The substituent at the anomeric centre of aldopyranoses usually prefers an axial orientation, rather than the expected equatorial position. Thus, for example, in the equilibrium between α- and β-D-

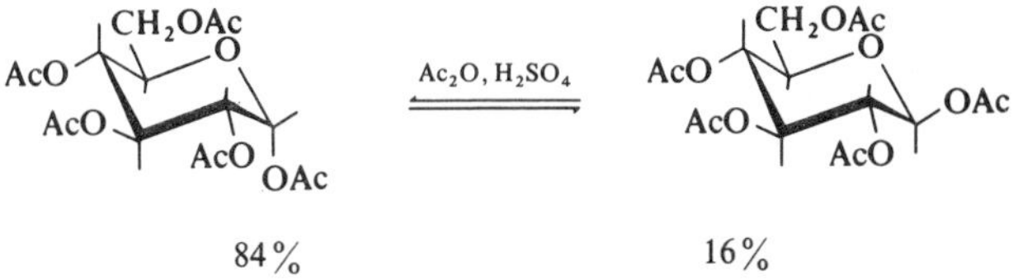

84%   16%

SCHEME 3.1

glucose penta-acetates (Scheme 3.1), the α (that is *ax*) C-1-acetate is the major component. This phenomenon, which is general, was termed the '*anomeric effect*' by Lemieux (1958) and it has been shown to decrease down the series halogen $> OCOPh > OCOCH_3 >$ OMe $>$ OH $> NH_2$. The effect is so powerful with halogen that, for example, the preferred conformation for 2,3,4-tri-*O*-acetyl-β-D-xylopyranosyl fluoride **3.18** is the all *ax* form **3.19** in which the

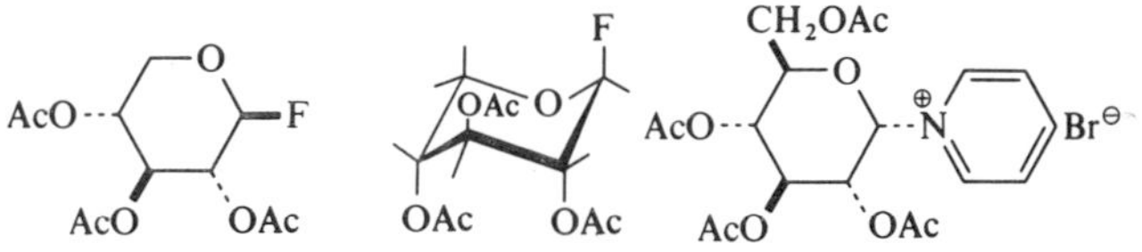

3.18   **3.19**   **3.20**

anomeric effect of the C-1-F group overcomes the effects of the three *ax*-acetate groups (Hall 1967) (see also p. 99).

Without doubt the anomeric effect is polar in nature, but the exact origin of the effect has yet to be elucidated.

From the above discussion a *reverse anomeric effect* could be predicted, and has indeed been observed in a limited number of cases. For example, the α-*gluco*-pyridinium compound **3.20**, exists in a boat-like conformation with the heterocyclic group equatorial, rather than the expected $^4C_1$ conformation in which it would be axial.

Furanose systems

The five-membered furanose ring has been much less studied than the pyranose system. Most studies that have been carried out show that one carbon of the ring is out of the plane of the other four. In nucleosides this is generally C-2 or C-3.

Acyclic systems

There has been much work recently on the conformation of open chain sugars and polyols, both by X-ray methods on crystals (Jeffrey, 1970), and by ^{1}H n.m.r. for solutions (Horton, 1965). A planar zig-zag conformation of the carbon backbone is generally observed, unless there is interaction between bulky groups in a 1,3-relationship. Such interactions cause deviations from the linear zig-zag conformation. Thus, for example, D-mannitol has the conformation **3.21**, with no 1,3-interactions; D-glucitol, which would have conformation **3.22** as a zig-zag, in fact has the bent conformation **3.23** because of the interaction between *O*-2 and *O*-4 in **3.22**.

3.21   3.22   3.23

It cannot be stressed too much that the detailed conformations of carbohydrate molecules must be taken into account whenever any reactions are considered.

4 Glycosides †

Synthesis

By alcoholysis

TREATMENT of a simple aldehyde or ketone with an alcohol in the presence of acid results in the formation of an acetal, by way of a hemiacetal (Scheme 4.1).

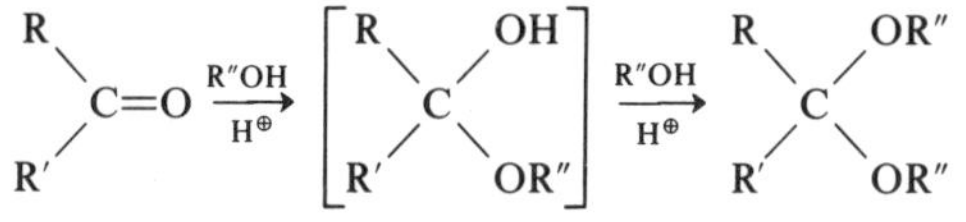

$$
\underset{R'}{\overset{R}{\diagdown}}C=O \quad \xrightarrow[H^{\oplus}]{R''OH} \quad \left[\underset{R'}{\overset{R}{\diagdown}}C\underset{OR''}{\overset{OH}{\diagup}}\right] \quad \xrightarrow[H^{\oplus}]{R''OH} \quad \underset{R'}{\overset{R}{\diagdown}}C\underset{OR''}{\overset{OR''}{\diagup}}
$$

Hemiacetal   Acetal

SCHEME 4.1

Under these conditions an aldose or ketose (which is a hemiacetal in the cyclic form) yields a cyclic mixed acetal, called a *glycoside*, shown diagrammatically in Scheme 4.2. The group R is known as the *aglycone*. Glycosides can have a furanose ring (furanosides), a pyranose ring (pyranosides) or rarely a septanose ring (septanosides).

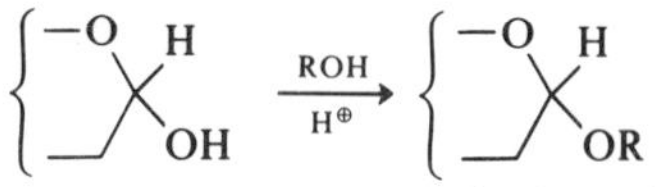

$$
\left\{\underset{\diagdown}{\overset{-O}{\diagup}}\right\}\underset{OH}{\overset{H}{\diagup}} \quad \xrightarrow[H^{\oplus}]{ROH} \quad \left\{\underset{\diagdown}{\overset{-O}{\diagup}}\right\}\underset{OR}{\overset{H}{\diagup}}
$$

Hemiacetal   Mixed acetal
ring form   or glycoside

SCHEME 4.2

For each ring size there will be two anomeric glycosides, α and β. In the complete description of a glycoside all these features must be included. Thus **4.1**, prepared by reaction of D-glucose with ethanol in the presence of hydrogen chloride is ethyl α-D-glucopyranoside.

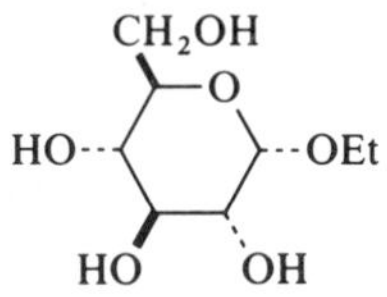

CH$_2$OH

HO---   ---OEt

HO   OH

4.1

† Internal glycosides (glycosans) are considered in Chapter 5.

The reaction of an aldose with an alcohol in the presence of anhydrous hydrogen chloride gives rise to a mixture of products, the composition of which varies with the aldose and the reaction conditions used. Thus D-glucose when reacted with hot 3% methanolic hydrogen chloride gives a mixture that is mainly comprised of pyranosides, of which the α-anomer preponderates. Reaction of the same sugar in the cold with methanol containing less than 1% hydrogen chloride gives mostly the α- and β-furanosides.

The mechanism of the reaction has been studied by various methods. Bishop (1962) quenched the reaction mixture at various times and examined the product by gas-liquid chromatography after the formation of TMS derivatives (see p. 32). It was found that the initially-formed furanosides were kinetically-controlled products, and these were slowly transformed to the thermodynamically-controlled products, the pyranosides. The reaction is more complex than this simple statement, because of the concurrent anomerisations between the α- and β- forms of each ring size. The reaction of D-xylose with methanolic hydrogen chloride followed the scheme shown in Figure 4.1. Formation of open-chain diacetals, such as **4.2** for the

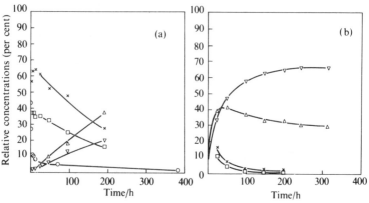

Fig. 4.1. Products from methanolysis of D-xylose (2% in 0·50% MeOH–HCl) at (a) 25° ±0·01°C, (b) 44° ±0·01°C. ○ D-xylose, × methyl β-D-xylofuranoside, □ methyl α-D-xylofuranoside, ▽ methyl α-D-xylopyranoside, △ methyl β-D-xylopyranoside. Reproduced by permission of the National Research Council of Canada from Bishop, C. F. and Cooper, F. P. *Can. J. Chem.* **40**, 224–32 (1962).

above system, has been demonstrated, but they comprise generally less than 2 per cent of the product mixture. It is not yet clear whether furanoside formation occurs via an acyclic ion, such as **4.3** or an

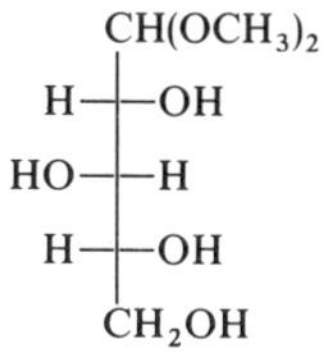

CH(OCH$_3$)$_2$

H—OH
HO—H
H—OH

CH$_2$OH

4.2

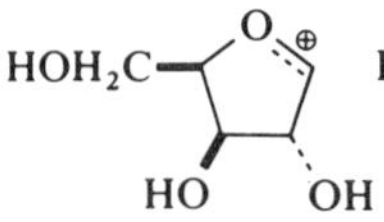

4.3 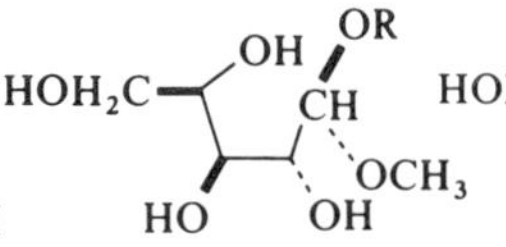   **4.4** R = H   **4.6**
         4.5 R = CH$_3$

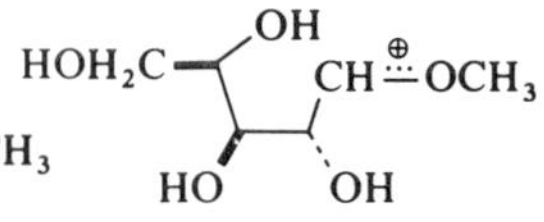

acyclic intermediate such as **4.4**. Similarly, furanoside to pyranoside interconversion may involve the dimethyl acetal **4.5** or an acyclic ion such as **4.6**. It is not certain of course that each individual aldose follows exactly the same reaction pathway.

From glycosyl halides

The Koenigs–Knorr method (1901) for the synthesis of glycosides is much used, yet still not completely understood. A fully acylated glycosyl bromide or the chloride, for example α-acetobromoglucose **4.7**, is treated with an alcohol or phenol under anhydrous conditions

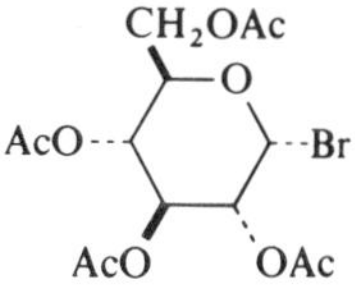

4.7

in the presence of a silver or mercuric salt. Glycosyl halides with a participating group, such as an acetate ester group at C-2, *trans* to the halide react *via* a 1,2-acyloxonium ion (*cf*. p. 41), that is with retention of configuration at C-1. When the potentially participating group and the C-1-halide are *cis*, then displacement occurs directly by an S_N1-type process (participation stereochemically impossible), unless anomerization of the halide is occurring concurrently, when there will be participation. In the above cases, then, the products are all *trans*-1,2-glycosides.

The synthesis of *cis*-1,2-glycosides by the Koenigs–Knorr method presents greater difficulties, and necessitates the use of a non-participating group at C-2, and a β-halide. Such an approach has been used by Wolfrom (1961) by using a 2-nitrate (see p. 81). Another non-participating group that has been used is the benzyl ether group (Fletcher 1969; Baddiley 1964).

From orthoesters

Recent work, mainly by Kochetkov and his group (1967 on) has demonstrated that orthoesters are useful precursors for the synthesis of 1,2-*trans*-glycosides. For example, treatment of **4.8** with an alcohol in nitromethane in the presence of mercuric bromide gives the β-glycoside **4.9**. A variation of the method, which is usually more

4.8   **4.9**   **4.10**

successful, is to convert the starting orthoester to the orthoester of the required alcohol **4.10** by exchange, and then to rearrange that to the required product **4.9**.

From glycals

Use of these starting materials has led to the most recent developments in the synthesis of α-glycosides. Addition to tri-*O*-acetyl-D-glucal **4.11** (see p. 64) of nitrosyl chloride gives a dimeric adduct **4.12**, which reacts with alcohols to give the α-glucoside 2-oxime **4.13** (Lemieux, 1968). The reaction proceeds *via* stereospecific addition to the 2-nitrosoglycal **4.14**.

The oxime **4.13** is a useful intermediate; choice of reducing agent can lead to either of the 2-amino-2 deoxy epimers (*gluco* or *manno*). Further, de-oximation of **4.13**, with, for example levulinic acid, and reduction of the carbonyl derivative thus formed with sodium borohydride gives the α-glucoside. These glycal-based reactions are useful for the synthesis of the important α-glycosidically-linked antibiotics (see p. 75).

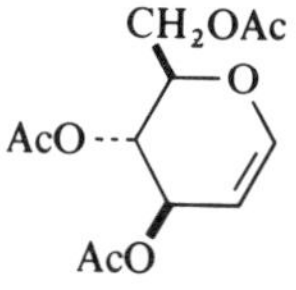

4.11 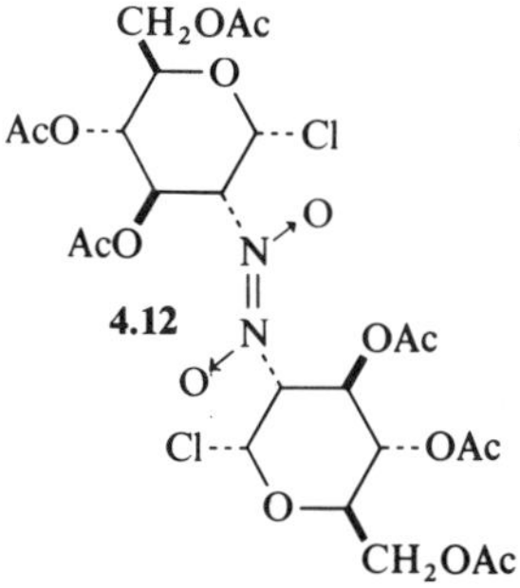   **4.12** 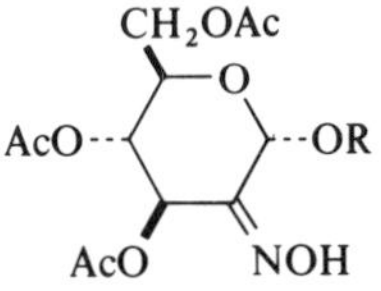   **4.13**

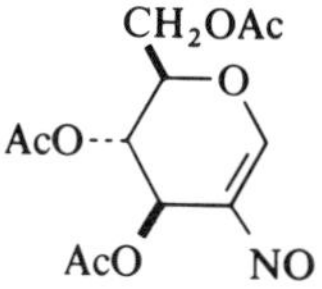

4.14

The use of glycals in the synthesis of 2,3-unsaturated α-glycosides is discussed on p. 64.

Hydrolysis

Because they are acetals, glycosides are very readily hydrolysed by dilute aqueous acid to the parent sugar and the aglycone.

Two modes of fission are possible in glycoside hydrolysis: (*a*) glycosyl-oxygen fission or (*b*) oxygen-aglycone fission (see **4.15**).

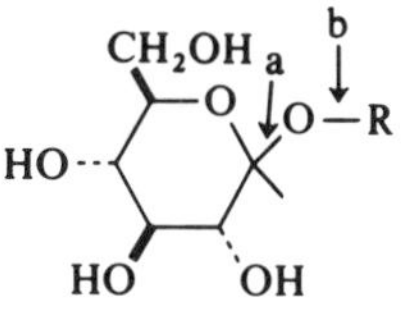

4.15 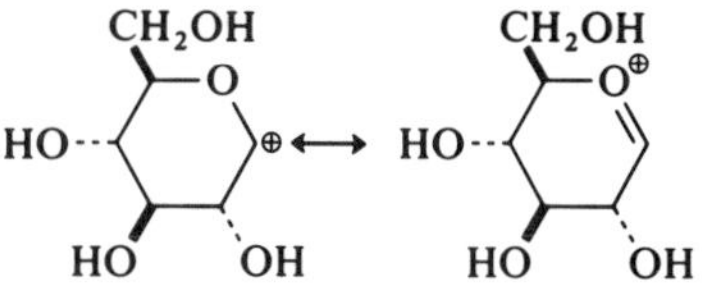   **4.16**

The mechanism was studied by using ^{18}O-enriched water as a hydrolysis medium (Bunton 1955). It was shown that route (*a*) was the usual mode of hydrolysis, proceeding *via* the acyclic carbonium ion **4.16**, which is then attacked by water to give the free sugar. A competing reaction is the attack on **4.16** of a free sugar molecule to give a disaccharide. This process is called 'reversion' (see also p. 85). Route (*b*) generally only occurs when the aglycone can form a more stable carbonium ion than **4.13**. An example is *tert*-butyl β-D-gluco-pyranoside **4.17**.

The group at C-2 can have a marked effect on the rate of hydrolysis. 2-Deoxy-glycosides are hydrolysed about one thousand times faster than the corresponding 2-hydroxy compounds. This effect is due

4.17

largely to removal of the inductive effect of the C-2-OH group (see p. 60 for C-2-NH$_2$ compounds).

Alkyl glycosides are hydrolysed by alkali only under drastic conditions, whereas some aryl glycosides are easily cleaved. The products are 1,6-anhydro-hexoses (see p. 34). Thus alkaline hydrolysis of phenyl β-D-glucopyranoside **4.18** gives 1,6-anhydro-β-D-

4.18   **4.19**

glucopyranose **4.19**; 1,2-epoxides are believed to be intermediates in this transformation. The *cis*-1,2-α-glucoside is recovered under the same reaction conditions.

5 Acetals and thioacetals

Acetals

ACETALS are generally derived from the reaction of two hydroxyl groups with the carbonyl group of an aldehyde or ketone, and thus have the general formula **5.1** or **5.2** depending on whether the

5.1   5.2   5.3

hydroxyl groups are in different molecules or in the same molecule (R^1 = H if the product is the derivative of an aldehyde) respectively. Type **5.2** are the most commonly encountered in carbohydrate chemistry. Carbohydrate compounds of type **5.1** may be of two types: one where the carbonyl group is part of a monosaccharide and the groups R^3 and R^4 are both non-carbohydrate, as in for example D-glucose dimethyl acetal **5.3**; and the other where one of the hydroxyl groups involved is part of a carbohydrate, as in a glycoside (see Chapter 4). Acetals are formed under anhydrous conditions in the presence of an acid catalyst. They are stable in alkaline conditions, but are readily hydrolysed by dilute aqueous acids (*cf.* the hydrolysis of glycosides, p. 23), and hence are very useful as blocking groups. A particularly useful reagent for hydrolysis of acetals is 90% aqueous trifluoroacetic acid used at room temperature (Goodman 1968).

Derivatives of single hydroxyl groups

Reaction of hydroxyl groups with dihydropyran in the presence of a trace of acid gives acetals such as **5.4** (*cf.* the reaction of glycals, p. 64) (Scheme 5.1). Such acetals find some use as blocking groups for single hydroxyl groups, but have the disadvantage that a diastereo-isomeric mixture is formed because of the creation of the new

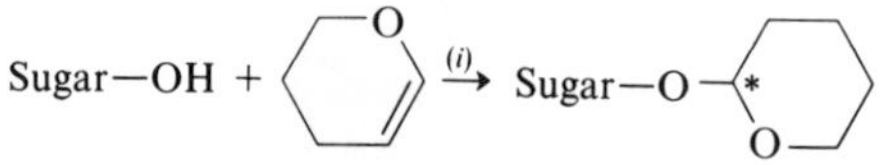

Sugar—OH + $\xrightarrow{(i)}$ Sugar—O—

(i) p-CH$_3$C$_6$H$_4$SO$_3$H

5.4

SCHEME 5.1

asymmetric centre (*). This can be overcome by the use of the dihydropyran derivative **5.5** (Reese 1967).

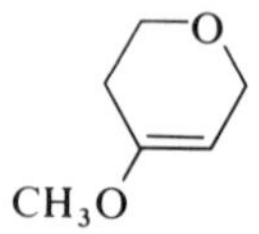

CH$_3$O

5.5

Derivatives of diols and carbonyl compounds

Ketones. The ketones most used are acetone and, less commonly, cyclohexanone. Ketones generally lead to the formation of five-membered acetal rings and the hydroxyl groups involved are usually *cis* (unless of course one of them is primary). Free sugars react in whichever form best accommodates this requirement. Thus, D-glucose when treated with acetone and sulphuric acid, reacts in the furanose form to give 1,2:5,6-di-*O*-isopropylidene-α-D-glucofuranose† **5.6** ('diacetone glucose'), whereas D-galactose reacts in the

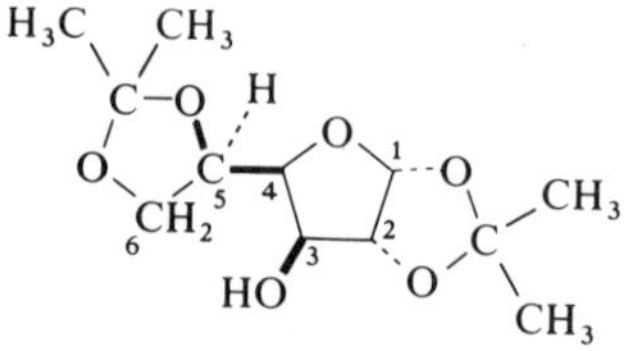

1,2:5,6-Di-*O*-isopropylidene-
α-D-glucofuranose

5.6

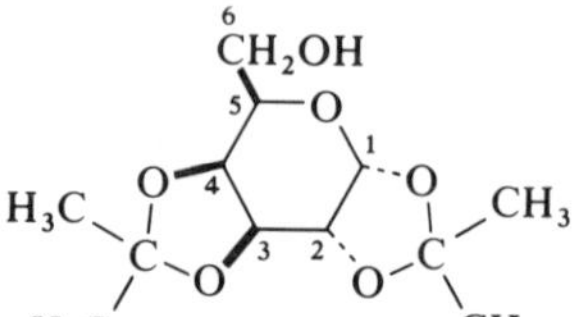

1,2:4,5-Di-*O*-isopropylidene-
α-D-galactopyranose

5.7

pyranose form to give 1,2:3,4-di-*O*-isopropylidene-D-galactopyranose **5.7**. When formation of a five-membered acetal ring is not possible, then a six-membered one may sometimes be formed. For

---

† In the nomenclature of acetals, the numbers of the hydroxyl groups bridged by each acetal group are separated by a comma, and each pair separated by a colon.

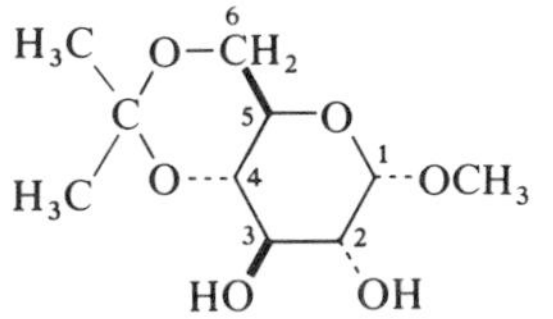

Methyl 4,6-*O*-isopropylidene-
α-D-glucopyranoside

5.8

example, methyl α-D-glucopyranoside gives the 4,6-*O*-isopropylidene derivative **5.8**.

Two *cis*-fused five-membered rings **5.9** are a relatively stable system and furanose di-acetals can generally be hydrolysed under very mild conditions to give compounds containing such a system. Thus, 1,2:5,6-di-*O*-isopropylidene-α-D-glucofuranose **5.6** yields 1,2-*O*-isopropylidene-α-D-glucofuranose **5.10**. With some sugars, more

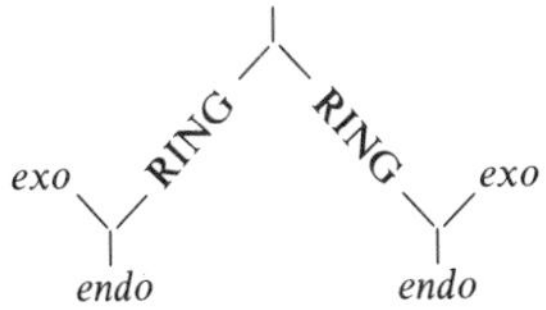

5.9

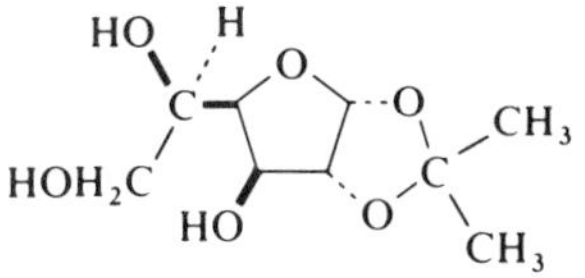

5.10

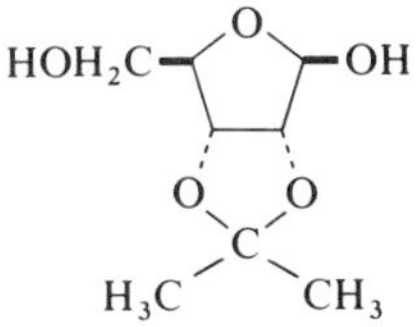

5.11

than one *cis*-fused five-membered ring system is possible. In these cases, the one with the minimum number of *endo* groups is formed. In theory, D-ribose could give α-1,2-, α-2,3- or β-2,3-*O*-isopropylidene acetals. The last one **5.11** has only one *endo* group (a methyl group) and is the one formed in practice.

Aldehydes. The commonly used aldehyde is benzaldehyde, and to a lesser extent, acetaldehyde and formaldehyde. Benzylidene derivatives are much used as blocking derivatives for the hydroxyl groups on C-4 and C-6 in pyranosides. Thus, methyl α-D-glucopyranoside on treatment with benzaldehyde in the presence of zinc chloride gives methyl 4,6-O-benzylidene-α-D-glucopyranoside **5.12**. Theoretically, two epimers should result because of introduction of a new asym-

Methyl 4,6-O-benzylidene-
α-D-glucopyranoside

5.12

metric carbon atom (∗). The formation is, however, an equilibrium process and the more stable isomer, namely that with the large phenyl group equatorially disposed on the acetal ring **5.12**, is formed preferentially. The 4,6-acetals of hexopyranosides may have *trans* or *cis* fused ring systems, analoges of *trans*- and *cis*-decalin respectively. The former, for example methyl 4,6-O-benzylidene-α-D-glucopyranoside **5.12**, have a rigid junction (*eq,eq*), since if both rings flipped to the opposite chair conformation the impossible *trans* (*ax, ax*) bridge would be necessary. The rigidity of the ring junction does not prevent the pyranose ring flipping to a boat or skew-boat conformation (see, for example, p. 36).

In *cis*-(*ax, eq*) systems, two isomeric products are possible on formation of the acetal, each derived from one chair conformation of the methyl hexopyranoside. In theory, therefore, methyl α-D-galactopyranoside could yield either **5.13** or **5.14** (the phenyl group is assumed to take up an equatorial position in both isomers). There is less non-bonded interaction in the 'O-inside' isomer **5.13** and this is preferred. Further, the 'O-inside' compound **5.13** can also exist in an alternative chair-chair conformation ('H-inside'), but the phenyl groups will then be axial.

(Pyranose ring D-$^4C_1$)

5.13

(Pyranose ring D-$^1C_4$)

5.14

As stated earlier one of the roles of acetal groups is the blocking of two hydroxyl groups. A useful reaction which unblocks and gives a functional group at C-6 is the reaction of a 4,6-*O*-benzylidene acetal with *N*-bromosuccinimide in carbon tetrachloride or similar solvent (Hannessian 1966, Hullar 1970). Generally the corresponding 6-bromo-4-*O*-benzoyl derivative is formed. In this way methyl 4,6-*O*-benzylidene-2,3-di-*O*-benzoyl-α-D-glucopyranoside **5.15** gives

Ph⟨O–CH₂, O, O⟩·· OCH₃   CH₂Br, O, BzO·· OCH₃ (Bz = PhCO)

BzO OBz   BzO OBz

5.15   **5.16**

methyl 6-bromo-2,3,4-tri-*O*-benzoyl-6-deoxy-α-D-glucopyranoside **5.16**. The bromo group is then available for other conversions, for example, into a 6-amino- or a 6-deoxy-sugar.

Thioacetals

In Chapter 4 the acid-catalysed reaction of sugars with alcohols was shown to give cyclic acetals or glycosides. Thiols (mercaptans), such as ethanethiol in the presence of concentrated hydrochloric acid, react to give derivatives of the aldehyde (acyclic) form of sugars. In this way, D-glucose gives the diethyl dithioacetal **5.17**. Treatment

EtS   SEt
   CH
H—OH   H—OCH₃
HO—H   CH₃O—H
H—OH   H—OCH₃
H—OH   H—OCH₃
CH₂OH   CH₂OCH₃

CHO (top right)

5.17   **5.18**

of such compounds with methanol in the presence of mercuric chloride and oxide gives methyl α- and β-furanosides, one of the better methods for their preparation. Thioacetals of ketoses are not obtained directly.

Fully methylated or acetylated dithioacetals can be treated with aqueous mercuric chloride to give derivatives of the open-chain aldehyde form of the sugar. Thus, methylation of **5.17** and removal of the acetal group gives 2,3,4,5,6-penta-*O*-methyl-*aldehydo*-D-glucose **5.18**.

6 Ethers and anhydro-sugars

THE formation of two types of ether are possible from the poly-hydroxylic systems of monosaccharides: (1) conventional *ethers*, formed with some other hydrocarbon residue, such as methyl ethers; and (2), internal or intramolecular ethers, called *anhydro-sugars*.

Ethers

Methyl ethers (ROMe)

These are the most common ethers in carbohydrate chemistry; some occur naturally. Before the advent of physical methods for structural study they were much used in the investigation of all types of carbohydrates from monosaccharides to polysaccharides.

The original methylation conditions are due to Purdie and Irvine (1903) and consist of refluxing the sugar derivative with methyl iodide in the presence of silver oxide. In general, several treatments are necessary before complete etherification, which is of course essential in structural work. Addition of NN-dimethylformamide (DMF) to the reaction mixture avoids the need for repeated treatments (R. Kuhn, 1955). A superior reagent system of the same type is a mixture of methyl iodide, DMF, and both the oxide and hydroxide of barium or strontium (R. Kuhn, 1961).

Methyl ethers can also be prepared by reaction of the carbohydrate derivative with dimethyl sulphate and aqueous sodium hydroxide (Haworth, 1915). Another method for methyl (and other alkyl) ethers is treatment of the carbohydrate derivative with sodium hydride in an aprotic solvent such as DMF followed by the addition of the alkyl iodide or bromide (Scheme 6.1).

$$\text{Sugar—OH} \xrightarrow{(i)} \text{Sugar—O}^{\ominus} \xrightarrow{(ii)} \text{Sugar—OCH}_3$$

(*i*), NaH, DMF; (*ii*) MeI

SCHEME 6.1

Diazomethane and boron trifluoride etherate can be used for methylation of sugar derivatives already bearing a base-labile substituent (Deferrari, 1967).

Methyl ethers are more volatile than the parent hydroxyl compound, and can be purified by high vacuum distillation. They have been used for gas–liquid chromatography and mass spectral analysis of carbohydrates because of this volatility. They were used in structural work because of their chemical stability, and because in general they do not migrate. If necessary methyl ethers can be cleaved by use of boron trichloride (Bourne, 1958) or tribromide.

Trimethylsilyl ethers ($ROSiMe_3$)

These compounds, frequently referred to as TMS derivatives, are much used in gas–liquid chromatography because of their ease of formation and great volatility when compared with the parent hydroxyl compound. Several silylating systems are available, a widely used one being a mixture of hexa-methyldisilazane ($Me_3SiNHSiMe_3$) and trimethylchlorosilane in pyridine (Sweeley, 1963). Another powerful reagent is bistrimethylsilylacetamide [$MeC(OSiMe_3)=NSiMe_3$]. The parent hydroxyl compound can be recovered from the TMS derivative by refluxing with 50% aqueous methanol.

Benzyl ethers ($ROCH_2Ph$, *often abbreviated to* R-OBn)

These are prepared in a similar fashion to methyl ethers, that is treating the sugar derivative with benzyl chloride in the presence of either silver oxide and DMF, or aqueous sodium hydroxide; alternatively the sodium hydride method can be used. Benzyl ethers have similar stability to methyl ethers, except that debenzylation to the parent hydroxyl compound is readily achieved by catalytic hydrogenation. For this reason, they are used quite widely as blocking groups in synthetic sequences.

Triphenylmethyl ('*trityl*') *ethers* ($ROCPh_3$, *often abbreviated to* R-OTr)

These derivatives are prepared by treatment of a sugar derivative with triphenylmethyl ('trityl') chloride in pyridine. The reaction is generally selective for primary hydroxyl groups; a few cases of the reaction of secondary trityl groups are known, but their formation is much slower. For example, methyl α-D-glucopyranoside gives the 6-*O*-trityl ether **6.1**.

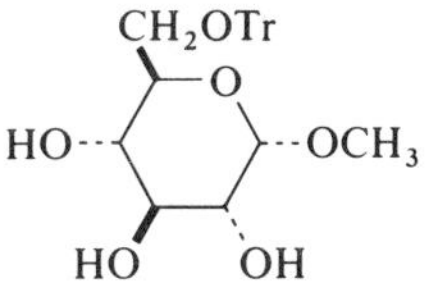

CH$_2$OTr

HO··· ⟩···OCH$_3$

HO OH

6.1

Trityl groups are rapidly and quantitatively removed at room temperature by hydrogen bromide in glacial acetic acid. Because of their selectivity for primary hydroxyl groups and their subsequent ready removal, trityl ethers are often used as blocking groups in synthetic sequences.

Allyl ethers (R.OCH$_2$.CH:CH$_2$)

These ethers are also useful as blocking groups. They can be prepared by the methods described for methyl and benzyl ethers and are stable to the majority of reagents used in carbohydrate chemistry. When required they can be removed by isomerization with potassium *t*-butoxide, to the corresponding propenyl ether which is cleaved to the parent hydroxyl compound with mercuric chloride and mercuric oxide in aqueous acetone (Gigg, 1967) (Scheme 6.2).

Sugar—OCH$_2$·CH:CH$_2$ $\xrightarrow{(i)}$ Sugar—O·CH:CH·CH$_3$ $\xrightarrow{(ii)}$ Sugar—OH

(*i*), KOCMe$_3$, DMSO; (*ii*) HgCl$_2$, HgO, aqMe$_2$CO

SCHEME 6.2

Anhydro-sugars

Anhydro-sugars can be considered to be formally derived from a monosaccharide, or a monosaccharide derivative, by the elimination of water from a pair of hydroxyl groups. Because of the polyhydroxylic nature of carbohydrates it is obvious that many types of anhydro derivative are possible. However, they can be divided into two fundamental groups: (1) *glycosans*, in which one of the two hydroxyl groups involved is at the anomeric centre, and (2) *internal ethers*, where both hydroxyl groups involved are alcoholic in character.

Glycosans

Since the anomeric hydroxyl group is involved in formation of the anhydro linkage, glycosans are strictly internal glycosides;

consequently they are readily opened by acidic hydrolysis. The most widely studied glycosans are the 1,6-anhydrohexopyranoses.

1,6-*Anhydro-aldohexoses*. Treatment of acetohalogeno-aldohexoses (for example acetobromoglucose **6.2**), with trimethylamine followed by barium hydroxide gives 1,6-anhydro-aldohexoses (1,6-anhydro-β-D-glucose **6.3**) (Karrer, 1921). They can also be prepared by pyrolysis of polysaccharides: cellulose yields 1,6-anhydro-β-D-glucopyranose (also called laevoglucosan) **6.3** and ivory nut mannan gives the corresponding mannose derivative. The structure of **6.3** and of the other 1,6-anhydro-aldohexoses was readily established since methylation followed by acid hydrolysis gave the 2,3,4-tri-*O*-methylaldose.

6.2   **6.3**   **6.4**

In a 1,6-anhydro derivative the aldohexose is in the D-$^1C_4$ (or L-$^4C_1$ conformation (see **6.3**), and the groups at C-1 and C-6 are both axial. In very dilute aqueous acid solution an equilibrium is set up between the aldohexose and its 1,6-anhydro derivative, the position of which is a reflection of the ease of attainment of the required conformation by a particular aldohexose. The percentage of anhydro-sugar present is given in Table 6.1 (Richtmyer, 1958,

TABLE 6.1
Equilibrium percentage of 1,6-anhydro-β-D-hexopyranoses at 100°C
(*Angyal*, 1968)

D-Hexose	Position of OH groups in $^1C_4$ conformation			Per cent anhydride
	C-2	C-3	C-4	
Idose	*eq*	*eq*	*eq*	86
Altrose	*eq*	*eq*	*ax*	65·5
Gulose	*ax*	*eq*	*eq*	65
Allose	*ax*	*eq*	*ax*	14
Talose	*eq*	*ax*	*eq*	2·8
Mannose	*eq*	*ax*	*ax*	0·8
Galactose	*ax*	*ax*	*eq*	0·8
Glucose	*ax*	*ax*	*ax*	0·2

Angyal, 1968). As would be predicted, the figures show that sugars such as D-glucose and D-mannose yield little anhydro-sugar (note that **6.3** has three axial hydroxyl groups), whereas ones such as D-idose and D-altrose give reasonable amounts of 1,6-anhydro compound (note that 1,6-anhydro-β-D-idopyranose **6.4** has three equatorial hydroxyl groups).

Internal ethers (Epoxides)

Three, four five, and six-membered internal ethers are known. The mode of formation is generally the same for all ring-sizes, namely the displacement of a good leaving group (L), generally a mesyloxy or tosyloxy group (see p. 43) by a suitably placed hydroxyl anion, (Scheme 6.3). The displacement is an S_N2 process so that unless

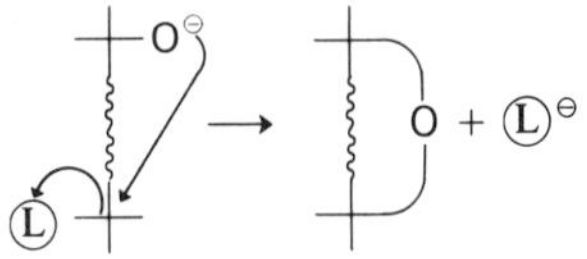

SCHEME 6.3

the leaving group is at a primary position, inversion occurs at the carbon atom involved.

Oxirans. These are the most important sub-class and are colloquially, but incorrectly called epoxides, which is strictly the name for the whole class of internal ethers. The general procedure for oxiran formation is the special case of Scheme 6.3 shown in Scheme 6.4. For such a reaction to occur the two groups involved must obviously

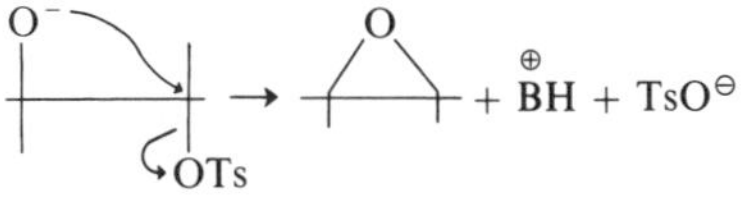

$Ts = p\text{-}CH_3C_6H_4SO_2$

SCHEME 6.4

be *trans*, and more important co-planar. On a pyranose ring this means that in the transition state, both groups must be axial, or essentially so. Thus, methyl 4,6-*O*-benzylidene-2-*O*-tosyl-α-D-glucopyranoside **6.5** yields the 2,3-anhydro-mannoside **6.6** on treatment with methanolic sodium methoxide, although in its ground state conformation **6.5** it has the reacting groups in *trans* diequatorial positions. The transition state must therefore approach the boat conformation **6.7** to fulfil the required stereochemistry. This is an example of a molecule reacting *via* an unfavourable conformation.

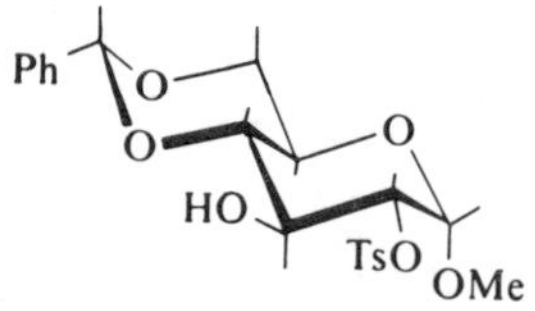

6.5 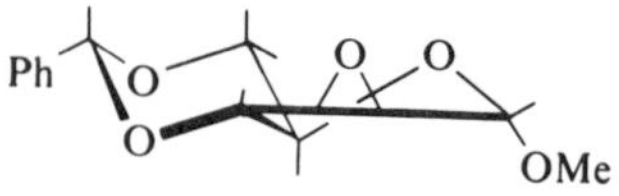 **6.6**

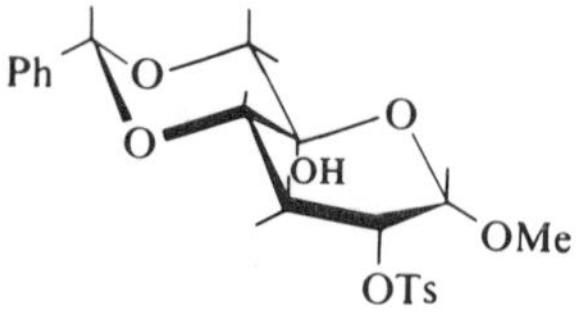

6.7 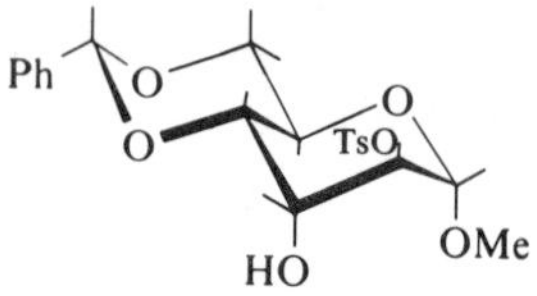 **6.8**

Similar boat intermediates are involved in the formation of many other epoxides. Methyl 4,6-*O*-benzylidene-2-*O*-tosyl-α-D-altropyranoside **6.8**, which has the reacting groups in axial positions in the ground-state conformation forms an epoxide extremely rapidly in the presence of a trace of base (Honeyman, 1958).

Oxirans can also be prepared by treatment of vicinal disulphonates with methanolic sodium methoxide. In this case one of the sulphonate groups undergo S_N2S attack at sulphur to give the required intermediate. It is not always possible to predict which of the two groups will undergo this cleavage. An example is the formation of the 2,3-anhydro-alloside **6.10** from methyl 4,6-*O*-benzylidene-2,3-di-*O*-tosyl-α-D-glucopyranoside **6.9**.

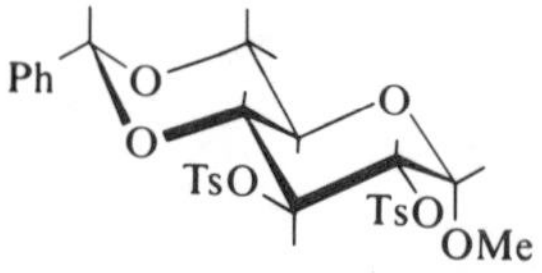

6.9 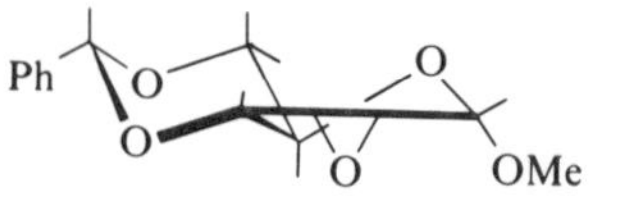 **6.10**

Oxirans may also be prepared by nitrous acid de-amination of suitably blocked *trans*-diaxial amino-sugars as shown in Scheme 6.5.

The importance of oxirans is in their use as intermediates in synthesis. The three-membered ring is readily opened by nucleo-

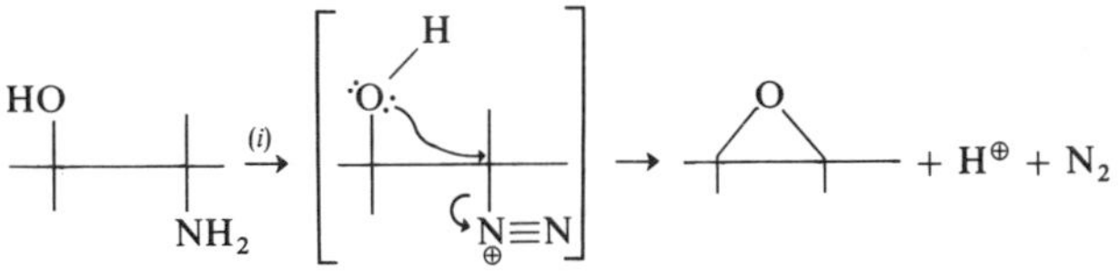

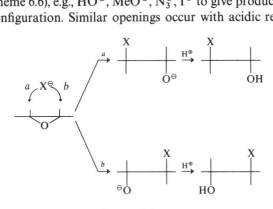

(i) NaNO$_2$, HCl

SCHEME 6.5

philes (Scheme 6.6), e.g., HO$^\ominus$, MeO$^\ominus$, N$_3^\ominus$, I$^\ominus$ to give products with a *trans* configuration. Similar openings occur with acidic reagents

SCHEME 6.6

such as hydrogen bromide though in this case the oxiran ring is first protonated and then opened. Pyranose oxirans can in theory exist in two conformations (for example **6.11** and **6.12**, for 2,3-oxirans)

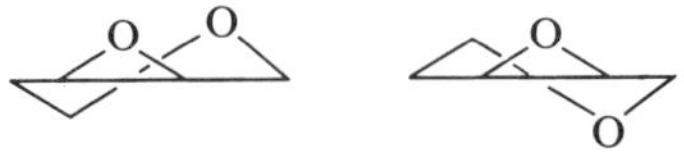

6.11   **6.12**

and so should therefore yield two products. This is generally the case for mobile systems, but for rigid pyranose epoxides such as the 2,3-anhydro-compounds **6.6** and **6.10**, the *trans* diaxial product is

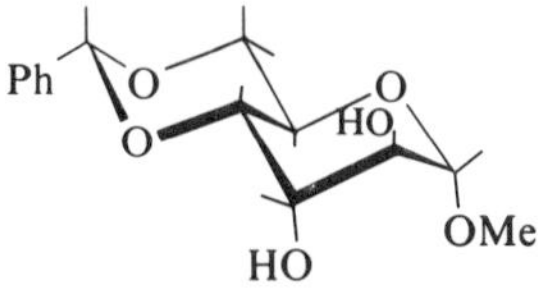

6.13

the major or sole product (Fürst–Plattner Rule: Mills, 1953).
Opening of the oxiran ring of **6.10** with hydroxide ion gives the
altroside **6.13** as the major product, providing a key step in the
synthesis of the rare sugar D-altrose from the common D-glucose.

6.14   **6.15**

3,6-*Anhydro-aldohexoses*. The internal ethers which can be prepared
from a suitably-blocked 3-hydroxy-6-O-tosyl (or mesyl) compound
exist in an extremely strained conformation if on a pyranoside ring,
that is **6.14**. Evidence for this strain comes from the ready rearrangement under acidic conditions to the furanoside form **6.15**.

7 Esters

As polyhydroxylic substances the monosaccharides (and their derivatives with free hydroxyl groups) are readily esterified by the processes used for simple alcohols; use of excess of reagent leads to compounds in which all of the available hydroxyl groups have reacted. Partial esterification can sometimes be achieved by using limited quantities of esterifying agent. Primary hydroxyl groups are, in general, more reactive than secondary ones and so these can be selectively esterified. For example, methyl α-D-glucopyranoside **7.1** can be converted into its 6-O-tosyl derivative **7.2** in good yield. The order of reactivity amongst secondary hydroxyl groups is not yet fully understood. For example, reaction of methyl 4,6-O-benzylidene-α-D-glucopyranoside **7.3** with one equivalent of tosyl chloride in pyridine yields the 2-ester with a very small amount of 2,3-diester, whereas under the same conditions, the β-anomer of **7.3** yields both 2- and 3-monoesters as well as the diester. Similar results are found with other reagents.

7.1 R = H

7.2 R = Ts

7.3

Esters are, of course, readily hydrolysed by alkali (though not always to the parent hydroxyl compound), but are more stable to acidic conditions.

Acetates

The usual reagent for acetylation is acetic anhydride in the presence of sodium acetate, zinc chloride, sulphuric or perchloric acids, or pyridine; the last named is the most commonly used. When the hydroxyl group on C-1 is free, either the α- or β-anomer may be formed. Here the influence of the catalyst is very important. For

7.4   7.5

example, D-glucose with acetic anhydride and zinc chloride gives
the α-penta-acetate **7.4**, whereas in the presence of sodium acetate
the β-anomer **7.5** is formed. Further the β-anomer **7.5** can be
anomerised to the α-form **7.4** on heating in acetic anhydride and
zinc chloride, or sulphuric acid (*cf.* p. 17).

De-acetylation can be accomplished very readily by using a basic
methanolic solution such as one containing sodium (Zemplen) or
barium methoxide, ammonia, or triethylamine.

It must be remembered that an anomeric acetoxy group will be
more reactive than the other acetoxy groups because it is a hemiacetal ester. The most useful of such selective reactions is that with
hydrogen bromide in acetic acid in which a C-1-bromo compound
is formed, a so-called acetobromoaldose. Thus, α- or β-D-glucose

7.6

pentacetate (**7.4** or **7.5**) gives acetobromoglucose (2,3,4,6-tetra-*O*-
acetyl-α-D-glucopyranosyl bromide) **7.6**. These reactive C-1 compounds are useful for the synthesis of a variety of derivatives, such
as glycosides (see p. 21) and glycals (see p. 64).

One important and often annoying reaction that can occur with
partially-substituted acetate esters is acetyl migration, usually under
extremely mild alkaline conditions. For example, the product from
the methylation with methyl iodide and silver oxide of methyl
2,3,4-tri-*O*-acetyl-β-D-glucopyranoside **7.7** is methyl 2,3,6-tri-*O*-
acetyl-4-*O*-methyl-β-D-glucopyranoside **7.8**, and not the expected
6-*O*-methyl ether **7.9**; that is, prior to methylation, the C-4-acetate
group migrated to C-6. The migration proceeds through the formation of a cyclic intermediate, in this case the essentially strainless

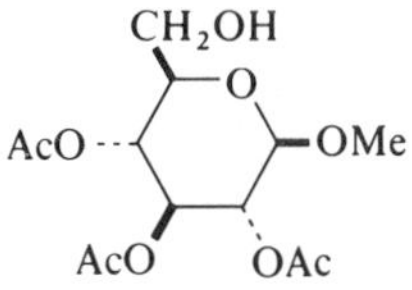

7.7 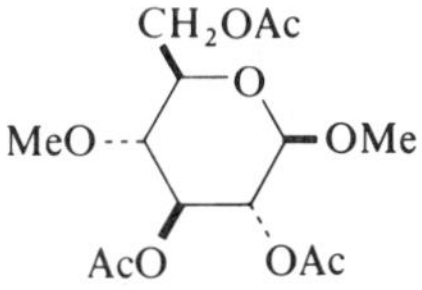   **7.8** 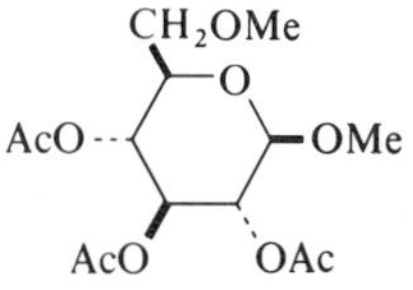   **7.9**

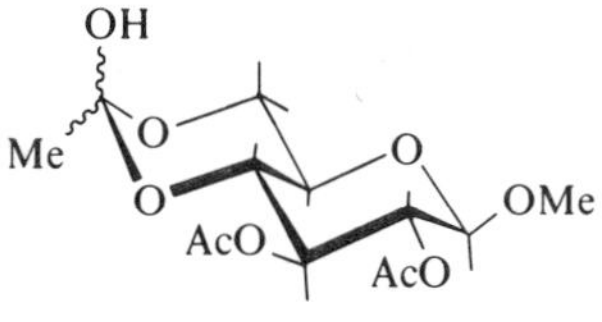

7.10

7.10 (cf. 4,6-*O*-benzylidene derivatives p. 28). The ease of migration, which in general takes place in the direction away from the anomeric centre, is related to the ease of formation of such a cyclic intermediate.

Reaction of polyacetates with antimony pentachloride in an organic solvent can cause estensive isomerizations *via* acetoxonium intermediates (Paulsen, 1967 on). Treatment of 2,3,4,6-tetra-*O*-acetyl-*β*-D-glucopyranosyl chloride **7.11** with antimony pentachlor-

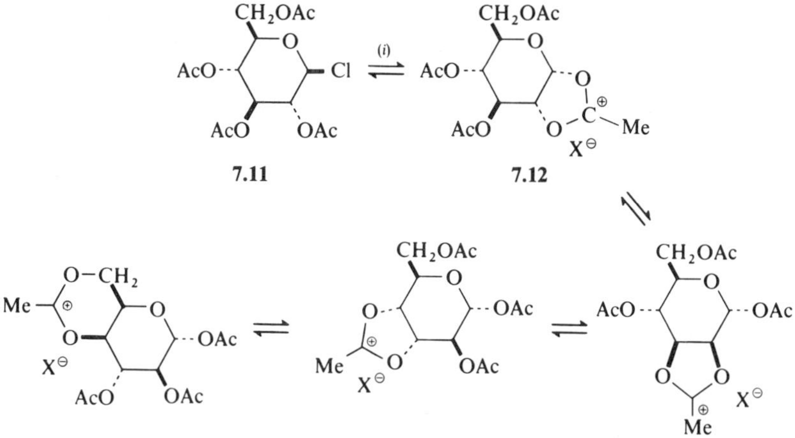

7.11   **7.12**

7.15   **7.14**   **7.13**

(*i*) SbCl$_5$: X = SbCl$_6$

SCHEME 7.1

ide in nitromethane causes an equilibrium to be set up between the acetoxonium ions **7.12** to **7.15** (Scheme 7.1), which are present in the ratio $60:12:7:21$. If methylene chloride is used as the solvent, the salt **7.15** crystallizes out and can be obtained in high yield. Hydrolysis and acetylation gives penta-O-acetyl-α-D-idopyranose **7.16**, providing a ready route for the synthesis of D-idose from D-glucose. It can be seen from Scheme 7.1 that the necessary steric requirement for the isomerization is the presence of an acetate *trans* next to the acetoxonium ion. For this reason the isomerizations that start from the D-galactose analogue of **7.11** can only involve two salts with the *galacto* **7.17** and *talo* **7.18** configurations: at equilibrium the ratio

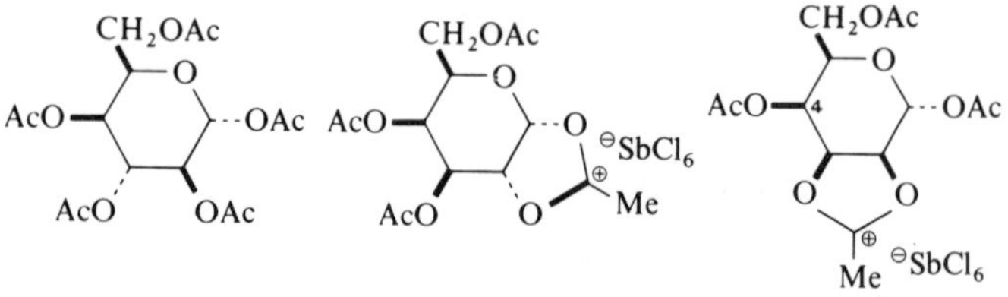

| **7.16** | **7.17** | **7.18** |

of **7.17**:**7.18** is $46:54$. The rearrangement cannot proceed further around the ring because in **7.18** the C-4-acetate group is *cis* to the 2,3-acetoxonium ion.

Chloroacetates are useful blocking groups as they can be removed by treatment with thiourea in hot methanolic benzene (Glaudemans, 1970) as in Scheme 7.2.

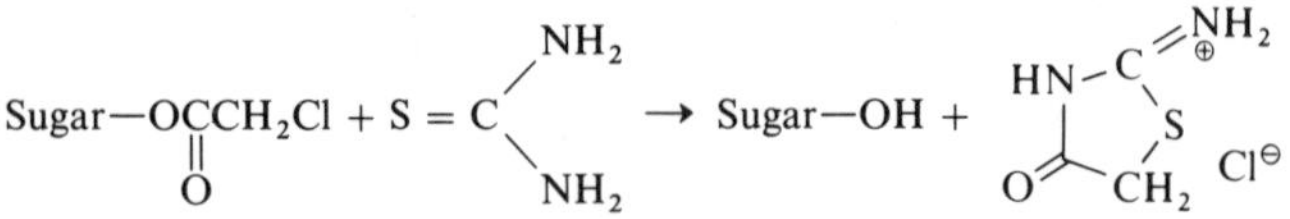

SCHEME 7.2

Benzoates

Benzoates are generally prepared from benzoyl chloride in pyridine. *p*-Nitro- and 3,5-dinitro-benzoates are sometimes used for characterization purposes as they are usually highly crystalline. Benzoates can migrate in the same way as acetates, and can also undergo similar rearrangements in the presence of antimony pentachloride *via* benzoxonium ions.

Selective benzoylation follows much the same pattern as that described for sulphonates above. *N*-Benzoylimidazole has been recommended as a selective benzoylating agent: for example compound **7.3** gives the 2-benzoate in 78 per cent yield.

Sulphonates

The commonest sulphonyl esters of sugars are toluene-*p*-sulphonates (*p*-tolysulphonates or 'tosylates', abbreviated to OTs) and the methanesulphonates ('mesylates', OMs). They are generally prepared from the sulphonyl chloride and the required carbohydrate in pyridine solution. For example, compound **7.3** gives **7.19**.

7.19

The importance of sulphonate esters in carbohydrate chemistry depends on their excellence as leaving groups in nucleophilic reactions, so that *via* such an ester, a hydroxyl group can be replaced by some other functional group. As expected, such displacement reactions occur most easily at primary positions, when solvents such as acetone or butan-2-one may be used. Displacements at secondary positions are much more difficult, but the use in recent years of aprotic solvents of high dielectric constant such as *NN*-dimethylformamide (HCONMe$_2$, DMF), dimethyl sulphoxide (Me$_2$SO, DMSO) and hexamethylphosphoramide [(Me$_2$N)$_3$PO, HMPT], has made such reactions common-place in carbohydrate chemistry. As expected they occur by an S_N2 process, one consequence of which is inversion of configuration at the reaction site. Commonly used nucleophiles are azide ion (N$_3^{\ominus}$) and hydrazine (to give 'blocked' amino-sugars), benzoate ion (to effect conversion to an epimer), and thiocyanate ion (desulphurization of the product gives a deoxy sugar). Like all S_N2 reactions, the ease of reaction is governed by steric factors. Dipolar interactions in the transition state between the incoming nucleophile and the departing sulphonyloxy group with other polar groups present in the molecule must be carefully considered (Richardson, 1968). The examples below illustrate the points made above.

(*a*) Displacement at C-4 in a glucopyranoside yields a galactopyranoside product (the C-4 epimer of the starting compound)

(Scheme 7.3).

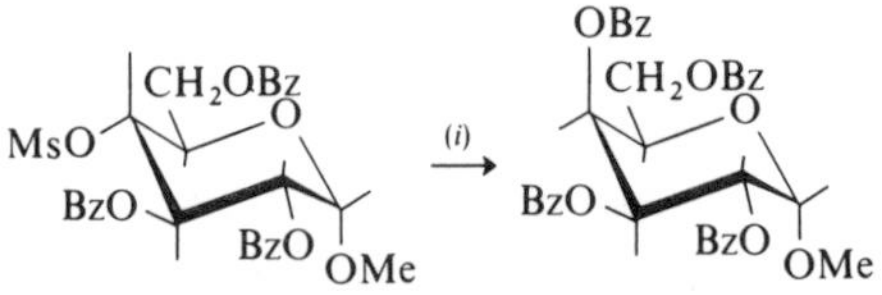

(i) NaOBz, DMF

SCHEME 7.3

(*b*) Comparison of displacements at C-3 in 1,2:5,6-di-*O*-isopro-pylidene-3-*O*-tosyl-α-D-glucopyranose **7.20** and in the corresponding allofuranose **7.21** show the difficulty of displacing the *exo*-sulphonyl-oxy group in **7.20** (cf. p. 27); the approach to the α-face in **7.20** is very hindered, especially to charged nucleophiles; hydrazine can enter under very forcing conditions. However, in the allose compound **7.21**, with an *endo*-sulphonyloxy group, the reaction is facile and a variety of nucleophiles can be used successfully to give 3-substituted glucose derivatives.

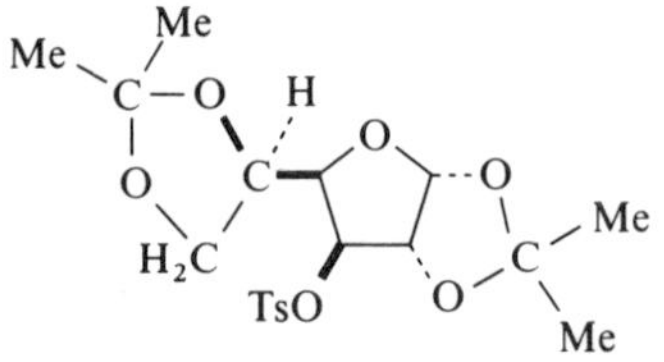

7.20 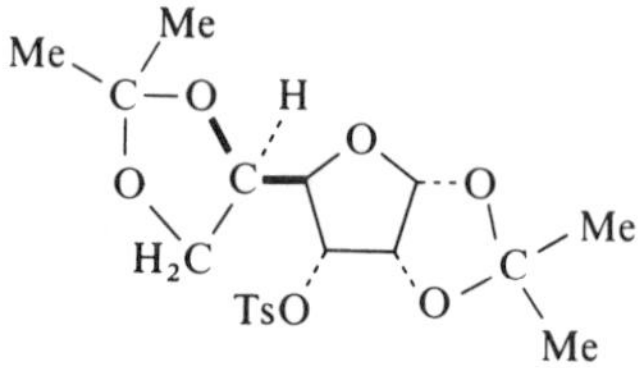   **7.21**

(*c*) Displacement at C-2 in aldose derivatives is generally impos-sible, because of the inductive effect of the adjacent anomeric acetal carbon atom, and because of steric interactions. For example con-sider the displacement in **7.22** viewed along the C-1, C-2 bond in **7.23** in the transition state. Note the dipolar interactions.

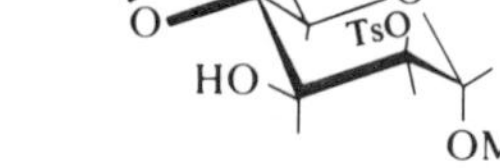

7.22

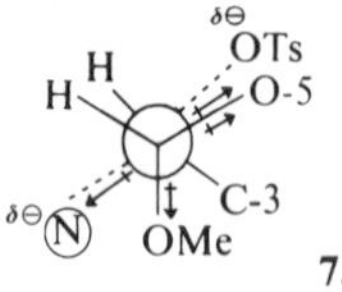

Newman projection along C-1, C-2 in transition state of displacement of TsO in **7.22**.

All the above reactions have involved displacement from an external nucleophile. Another major class of reactions of sulphonyloxy compounds involves intramolecular attack, that is *neighbouring group participation* (NGP). Such attack may be the only mode of displacement of a sulphonyloxy group as in the formation of oxirans from vicinal hydroxy-sulphonates (see p. 35), or may cause the reaction to pass through an intermediate which is then attacked by an external nucleophile. This latter type of NGP can compete with direct displacement of the sulphonyloxy group, depending on the nucleophile and the solvent involved. For *adjacent* centres the steric

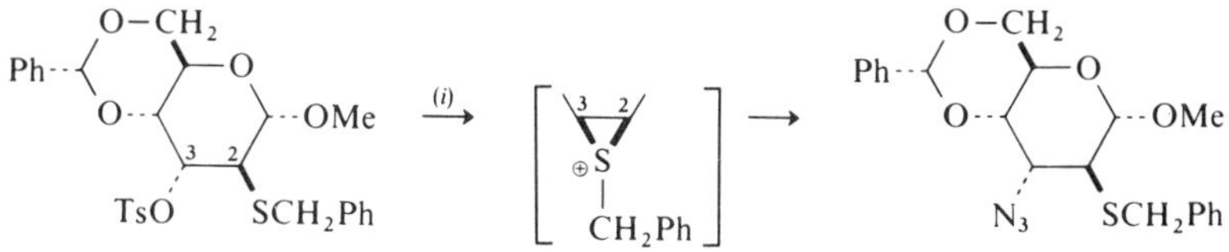

(*i*) NaN$_3$, DMF

SCHEME 7.4

requirement for effective NGP is that the participating group and the sulphonyloxy group must be antiparallel in the transition state. Schemes 7.4 and 7.5 illustrate some of these points (see also p. 35 and 59).

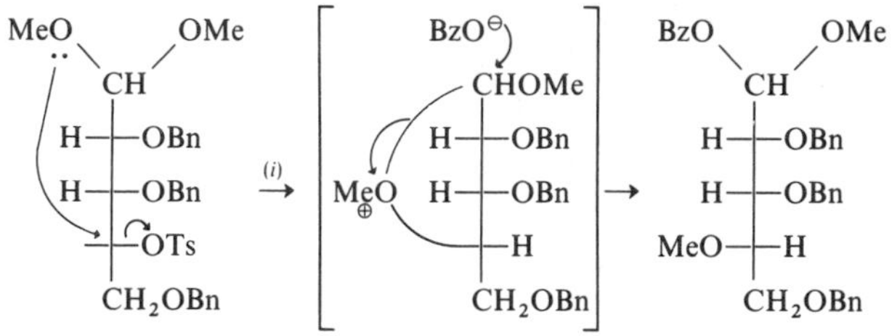

(*i*) NaOBz, DMF. Bn = CH$_2$Ph

SCHEME 7.5

In spite of the variety of reactions described sulphonate esters can generally be cleaved to the parent hydroxyl compound by use of sodium amalgam, lithium aluminium hydride (primary sulphonates yield the CH$_3$ compound not CH$_2$OH), by photolysis in the presence of methoxide ion, or by use of sodium–naphthalene reagent.

Phosphates

Carbohydrate phosphates are of particular interest because of their central role in biochemical processes, such as biosynthesis and sugar metabolism. In particular, phosphates of D-ribose and 2-deoxy-D-ribose are very important, as the basic components of the nucleic acids and many co-enzymes.

Aldosyl 1-phosphates are readily prepared from a fully acetylated aldosyl bromide and silver phosphate. The most general method for preparing phosphates at other positions is the use of a disubstituted phosphoryl monochloride, such as the diphenyl or dibenzyl derivative on the appropriately blocked sugar, followed by hydrogenation (Scheme 7.6).

$$\text{Sugar}-\text{OH} \xrightarrow{(i)} \text{Sugar}-\overset{\overset{\text{O}}{\|}}{\text{OP}}(\text{OCH}_2\text{Ph})_2 \xrightarrow{(ii)} \text{Sugar}-\overset{\overset{\text{O}}{\|}}{\text{OP}}(\text{OH})_2$$

$$(i)\ \text{Cl}\cdot\overset{\underset{\|}{\text{O}}}{\text{P}}(\text{OCH}_2\text{Ph})_2;\ (ii)\ \text{H}_2,\ \text{Pd}$$

SCHEME 7.6

Another reagent system, used particularly in nucleotide synthesis is cyanoethyl phosphate and dicyclohexylcarbodiimide (DCC). These form the intermediate **7.24** as shown which is then attacked by the sugar to give the phosphate derivative **7.25** (Scheme 7.7). The cyanoethyl group is readily removed from **7.25** under mild alkaline conditions by virtue of the labile proton α to the cyano group **7.26**.

Phosphate esters can migrate in the presence of dilute acids.

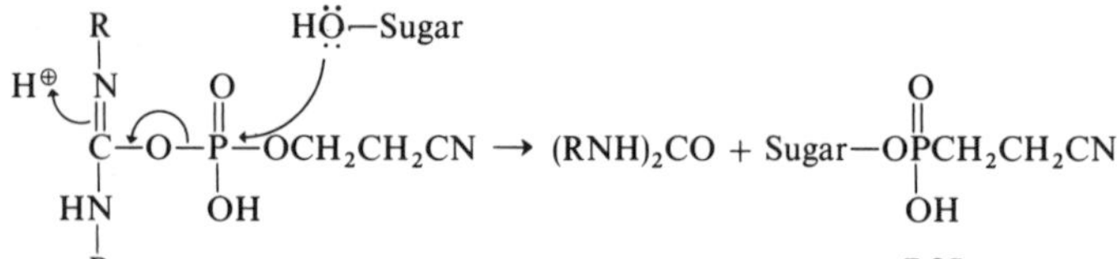

$$\text{H}^{\oplus}\overset{\text{R}}{\underset{\text{HN}}{\underset{|}{\overset{|}{\text{N}}}}}\overset{}{\underset{\underset{\text{R}}{|}}{\text{C}}}\text{-O-}\overset{\overset{\text{O}}{\|}}{\underset{\underset{\text{OH}}{|}}{\text{P}}}\text{-OCH}_2\text{CH}_2\text{CN} \rightarrow (\text{RNH})_2\text{CO} + \text{Sugar}-\overset{\overset{\text{O}}{\|}}{\underset{\underset{\text{OH}}{|}}{\text{OP}}}\text{CH}_2\text{CH}_2\text{CN}$$

HӦ—Sugar

7.24   **7.25**

R = Cyclohexyl

SCHEME 7.7

$$\text{Sugar}-\text{O}\overset{\overset{\text{O}}{\|}}{\underset{\underset{\text{HO}}{|}}{\text{PO}}}\text{-CH}_2\text{-CHCN}$$

H

OH

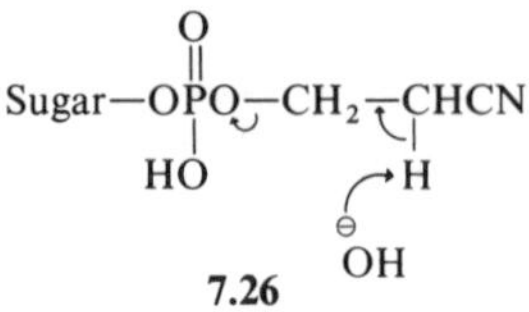

7.26

Miscellaneous esters

The above classes of ester are the most important, but others have found use in carbohydrate chemistry, some of which are described briefly below.

Carbonates. Either cyclic or acyclic carbonates can be formed depending on the reagent used and the number and stereochemistry of the hydroxyl groups involved. Phosgene ($COCl_2$) in pyridine gives only cyclic esters, which occupy the same positions as the rings in the corresponding isopropylidene acetals (see p. 26); for example D-glucose gives D-glucose 1,2:5,6-dicarbonate **7.27**. Partial alkaline hydrolysis of **7.27** gives (in contrast to the di-acetal), the 5,6-mono-carbonate a useful precursor in the synthesis of furanosides.

7.27   **7.28**

Alkyl chloroformates in the presence of aqueous sodium hydroxide give derivatives with cyclic and acyclic groups. Thus, methyl α-D-mannopyranoside gives **7.28** with ethyl chloroformate.

Thionocarbonates are useful precursors of olefinic sugars (see p. 65).

Sulphates. These can be prepared by the action of chlorosulphonic acid or sulphur trioxide in pyridine. Used little in monosaccharide chemistry, they are important in polysaccharide chemistry. Several sulphated polysaccharides are found naturally, for example, heparin, the natural blood anticoagulating agent, and many seaweed polysaccharides, such as the carrageenans.

Nitrates. These esters are best prepared by the use of fuming nitric acid in chloroform; an alternative method is with dinitrogen pentoxide in chloroform.

The behaviour of nitrate esters falls midway between that of carboxylates and sulphonates. For example, a primary nitrate can be displaced by iodide ion, whereas a secondary nitrate under the same conditions gives the parent hydroxyl compound.

Denitration of both primary and secondary nitrates to the hydroxyl compound can also be achieved by use of hydrazine.

Fully nitrated carbohydrates, such as those of glycerol and cellulose (incorrectly called 'nitroglycerol' and 'nitrocellulose') are used as explosives. Mannitol polynitrates have found use in the treatment of heart disease.

8 Oxidation reactions

Oxidation of single hydroxyl groups

Sugar acids

Oxidation of the potential aldehyde group of aldoses, or of the primary hydroxyl group generates carboxylic acid functions. Mild oxidation of an aldose by bromine water, for example, yields the corresponding *aldonic acid* in which only the aldehyde function is oxidized; the acid is usually isolated as a γ-lactone. For example D-glucose gives gluconic acid **8.1** usually isolated as D-glucono-γ-lactone **8.2**. More drastic oxidation, say by nitric acid or by hypo-

8.1   8.2   8.3

iodite yields a dicarboxylic acid, called a *saccharic* or more correctly an *aldaric acid*, usually isolated as a dilactone, such as **8.3** from D-glucose.

Acids with a carboxylic acid function only at C-6, the *uronic acids* can be prepared by oxidising a suitably blocked derivative with only the hydroxymethyl group free, such as **8.4**; potassium permanganate has been used as oxidant in such reactions. Free uronic acids are isolated as their γ-lactones in which the size of the lactol ring varies from sugar to sugar. D-Glucuronic acid yields the furanose derivative

8.4   8.5   8.6

8.5, with the favoured *cis*-fused five-membered ring arrangement, whereas D-galacturonic acid gives the pyranose derivative **8.6**, because the formation of a furanose form would need *trans*-fused five-membered rings. These compounds are analogous stereochemically to the 3,6-anhydrohexoses (see p. 38).

Uronic acids are found naturally as the components of several polysaccharides (see p. 94).

Alduloses

'Ulose' is the name-ending given to monosaccharide derivatives containing a ketonic carbonyl group. Thus **8.7** is an aldulose deriva-

$$CH_2OH$$

8.7

tive, more precisely it is methyl β-D-*arabino*-hexopyranosid-2-ulose.†
As in general organic chemistry, such carbonyl compounds are extremely useful synthetic intermediates: reduction may yield the epimer of the starting alcohol; Grignard and similar additions yield branched-chain sugars (p. 68); oximation and reduction of the product gives amino derivatives (p. 22).

Because of the polyhydroxylic nature of monosaccharides, most of the methods for the synthesis of alduloses require the use of a derivative in which all the hydroxyls but the one to be oxidized are blocked. The mildest oxidation methods are based on the use of dimethyl sulphoxide (DMSO). Pfitzner and Moffatt (1963) discovered that DMSO in conjunction with dicyclohexylcarbodiimide (DCC) and phosphoric acid or pyridinium trifluoroacetate, smoothly oxidizes isolated hydroxyl groups to the corresponding ketone (or aldehyde if a primary hydroxyl) in high yield. Acid anhydrides, such as acetic anhydride or phosphoric anhydride in DMSO have also been used successfully. The last reactions are mechanistically related to the Pfitzner–Moffatt method and also proceed *via* alkoxy-sulphoxium salts $[R_2CHO\overset{\oplus}{S}(CH_3)_2]$. These oxidations are sometimes accompanied by formation of the methylthiomethyl ether of the parent hydroxyl compound $(R_2CHOCH_2SCH_3)$.

† In naming compounds of this type, the configuration of the chiral centres is denoted by an italicized sugar name, here '*arabino*'. Note that although the compound is a hexose, this is a pentose prefix because there are only three chiral centres.

Baker (1965) noted that oxidation of some sugar derivatives with an axial substituent vicinal to the hydroxyl group, gave the equatorial epimer of the expected product. Applications of these oxidations are shown in Scheme 8.1.

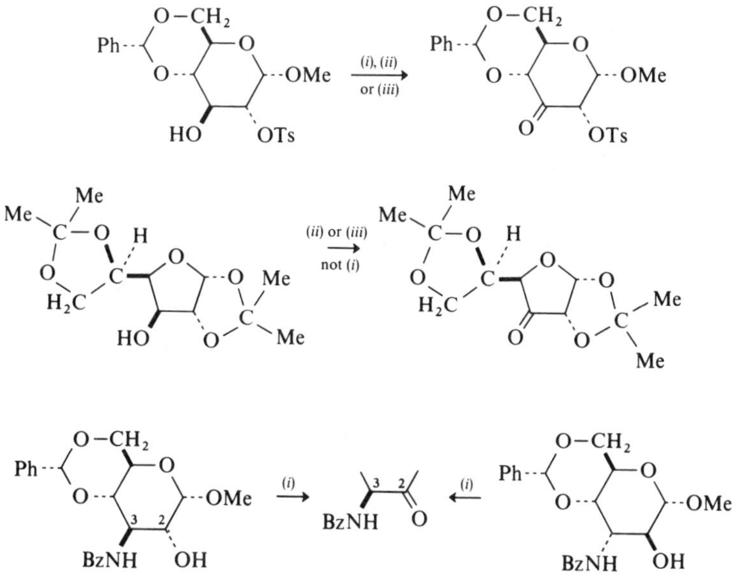

(i) DMSO, DCC; (ii) DMSO, Ac$_2$O; (iii) DMSO, P$_4$O$_{10}$

SCHEME 8.1

Another mild oxidant in current use is ruthenium tetroxide, conveniently prepared *in situ* from the dioxide and potassium periodate (Overend, 1968).

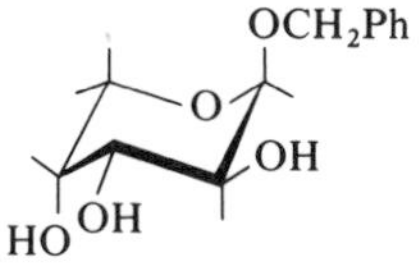

8.8 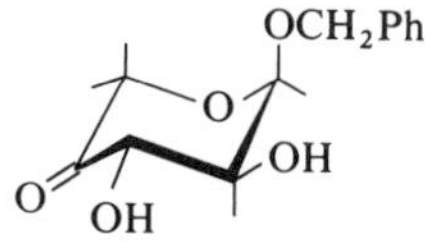   **8.9**

Selective oxidation of polyhydroxylic compounds occurs with platinum and oxygen in neutral solution (Heynes, 1962); axial hydroxyl groups are preferentially oxidized. Thus, benzyl β-D-arabinopyranoside **8.8** gives benzyl β-D-*threo*-pentosid-4-ulose **8.9**.

Reduction of aldulose derivatives with sodium borohydride can lead to the formation of the epimer of the starting hydroxyl compound. Routes to several less accessible sugars depend on this approach (see p. 109).

Oxidation of vicinal diols and amino-alcohols

Periodic acid and its sodium and potassium salts

The oxidation of vicinal diol (α-glycol) groups by sodium or potassium periodate was discovered by Malaprade (1928) and its specificity was shown by Fleury and Lange (1932). The oxidation, which is quantitative, is usually carried out in aqueous solution, using periodic acid for solutions of pH < 4, sodium metaperiodate for pH 4–7 and the potassium salt for alkaline oxidations; most oxidations are carried out at about pH 4. Light must be excluded to avoid decomposition of the reagent (Head 1950). Periodate uptake is measured by releasing iodine from the periodate ion and subsequent titration, or by spectroscopic methods. Methods have been developed for oxidation of microgram quantities of material.

In the oxidation of a vicinal diol, the carbon–carbon bond is broken and two carbonyl groups are formed (Scheme 8.2), aldehydes

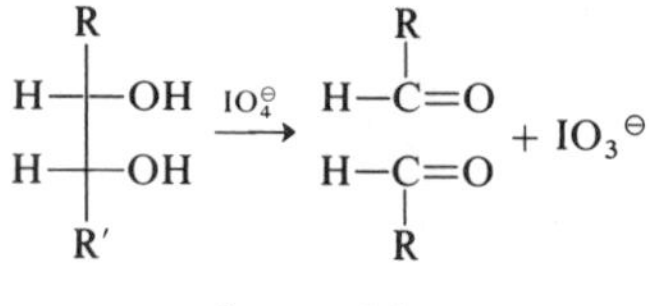

$$
\begin{array}{c}
\text{R} \\
\text{H}\!-\!\!\!-\!\text{OH} \\
\text{H}\!-\!\!\!-\!\text{OH} \\
\text{R}'
\end{array}
\;\xrightarrow{\;\text{IO}_4^{\ominus}\;}\;
\begin{array}{c}
\text{R} \\
\text{H}\!-\!\text{C}\!=\!\text{O} \\
\text{H}\!-\!\text{C}\!=\!\text{O} \\
\text{R}
\end{array}
\;+\;\text{IO}_3^{\ominus}
$$

Scheme 8.2

from secondary alcohols, ketones from tertiary ones. If one of the hydroxyl groups is primary, then formaldehyde is produced (Scheme

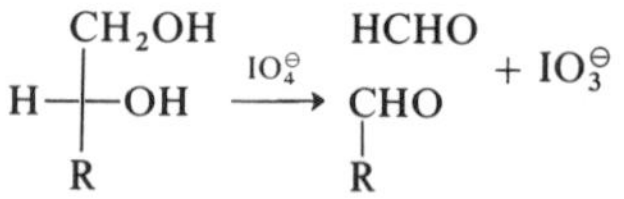

$$
\begin{array}{c}
\text{CH}_2\text{OH} \\
\text{H}\!-\!\!\!-\!\text{OH} \\
\text{R}
\end{array}
\;\xrightarrow{\;\text{IO}_4^{\ominus}\;}\;
\begin{array}{c}
\text{HCHO} \\
\text{CHO} \\
\text{R}
\end{array}
\;+\;\text{IO}_3^{\ominus}
$$

Scheme 8.3

8.3); this can be separately analysed and the number of primary groups present accurately determined.

Compounds containing three contiguous hydroxyl groups consume two proportions of periodate and yield two carbonyl groups

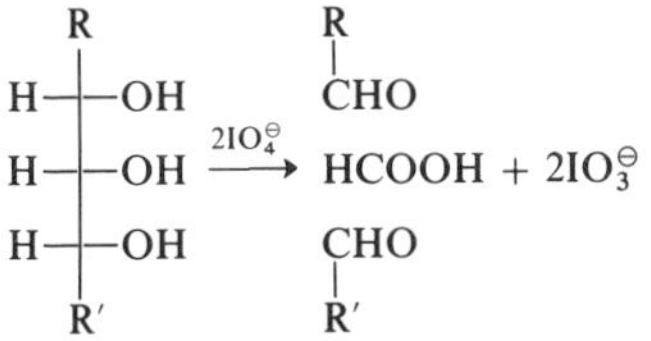

$$\begin{array}{c}
\text{R} \\
\text{H}\!-\!\!-\!\text{OH} \\
\text{H}\!-\!\!-\!\text{OH} \\
\text{H}\!-\!\!-\!\text{OH} \\
\text{R}'
\end{array}
\xrightarrow{\;2\text{IO}_4^{\ominus}\;}
\begin{array}{c}
\text{R} \\
\text{CHO} \\
\text{HCOOH} + 2\text{IO}_3^{\ominus} \\
\text{CHO} \\
\text{R}'
\end{array}$$

SCHEME 8.4

plus a carboxylic acid, usually formic acid as carbohydrates generally have secondary hydroxyl groups (Scheme 8.4). The formic acid produced can be analysed quantitatively by titration.

The oxidation is believed to occur *via* formation of a five-membered cyclic ester (Criegie 1933, Bunton 1954). For pyranosyl derivatives *ax,eq*-diols react faster than *eq,eq* ones, because of the greater ease of formation of the intermediate. However, rigid *ax,ax*-diols or other systems in which the hydroxyl groups are held essentially 180° apart are not oxidized, because the stereochemistry is impossible for cyclic intermediate formation. Examples are methyl 4,6-*O*-benzylidene-α-D-altroside **8.10** and 1,6-anhydro-β-D-glucofuranose (11).

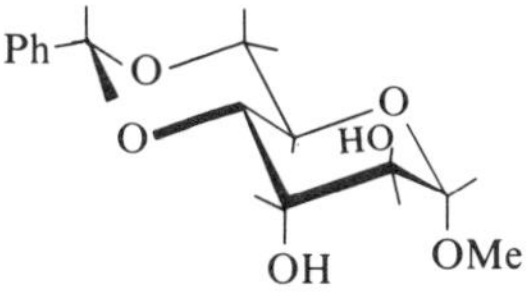

8.10 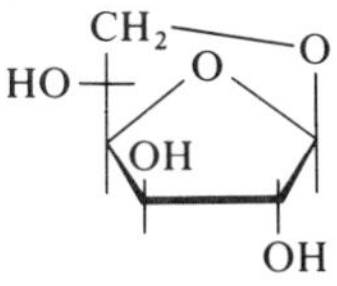   **8.11**

It can be seen, therefore, that periodate oxidation can be used to determine accurately the number, type, and arrangement of hydroxyl groups in a polyhydroxylic molecule. It has found extensive use in polysaccharide chemistry (see p. 85) and in the determination of the ring-size of aldohexosides (Hudson 1936): pyranosides reduced two moles of periodate, and released one mole of formic acid and no formaldehyde, whereas furanosides reduced two moles of oxidant also, but released one mole of formaldehyde and no formic acid. The dialdehydes from such oxidations have been isolated after oxidation and formation of barium or strontium salts (Hudson 1936) or by reduction with sodium borohydride (Smith 1955). Such di-aldehydes have also found use in the synthesis of amino-sugars (see p. 59).

Periodate oxidation of vicinal amino-alcohols in which the amino group is primary or secondary are oxidized in the same way as diols except that in addition to carbon–carbon bond cleavage, ammonia or an amine are also produced. Another difference is that rigid *ax,ax* systems are very slowly oxidized. Systems in which the amino group is tertiary are oxidized to the corresponding *N*-oxide (Guthrie 1971). Furanosides of 3-amino-3-deoxy-sugars are oxidized anomalously and so periodate uptake must be used with caution in structural studies on amino-sugars. Vicinal hydroxy-acylamido systems are not oxidized by periodate.

Lead tetra-acetate

Oxidations of vicinal diols with this reagent are similar to those with periodate, but benzene or glacial acetic is the usual solvent. The oxidation is also believed to occur *via* a five-membered cyclic intermediate (Criegie 1933). In pyridine rigid *ax,ax*-diol systems such as **8.10** are oxidized quite rapidly (Perlin 1960).

9 Nitrogen-containing monosaccharide derivatives

Glycosylamines

THE parent glycosylamines have an amino group in place of the anomeric hydroxyl group; derivatives of secondary or tertiary amines are named as *N*-substituted compounds. Thus **9.1** is β-D-xylopyranosylamine and **9.2** is *N*-phenyl-β-D-glucopyranosylamine. Compounds of the latter type, *N*-aryl-glycosylamines, are easy to prepare, by adding the monosaccharide to the amine in aqueous solution containing a trace of acid. As with free sugars, glycosylamines exist as an equilibrium mixture of acyclic and cyclic forms in solution.

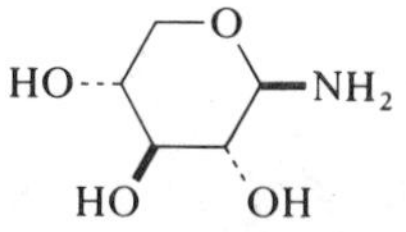

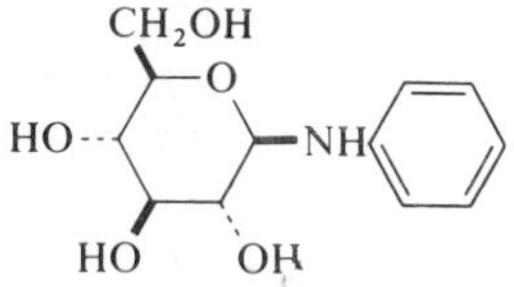

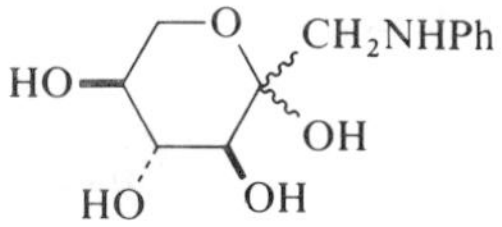

Glycosylamines rearrange in the presence of acid to 1-amino-1-deoxyketoses, the so-called Amadori rearrangement (Amadori 1925), for example **9.2** to 1-phenylamino-1-deoxy-D-fructose **9.3**. The pathway followed by this rearrangement (Weygand 1939) is shown in Scheme **9.1**.

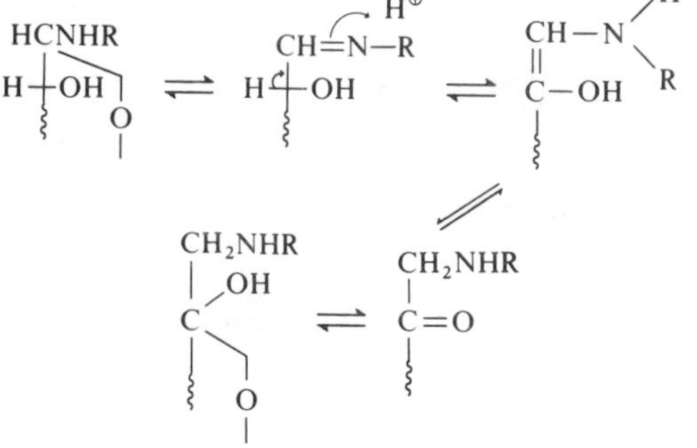

SCHEME 9.1

Nucleosides

Nucleosides, the important constituents of the nucleic acids, DNA and RNA, are compounds of the glycosylamine type in which the aglycone is a purine or pyrimidine base. The sugar component is always D-ribofuranose or 2-deoxy-D-ribofuranose (strictly 2-deoxy-D-*erythro*-pentofuranose), and the anomeric configurations in the natural compounds are almost invariably β.

The structures of the most commonly derived nucleosides from ribonucleic acid (RNA) are shown in Scheme 9.2 and those from deoxy-ribonucleic acid (DNA) in Scheme 9.3.

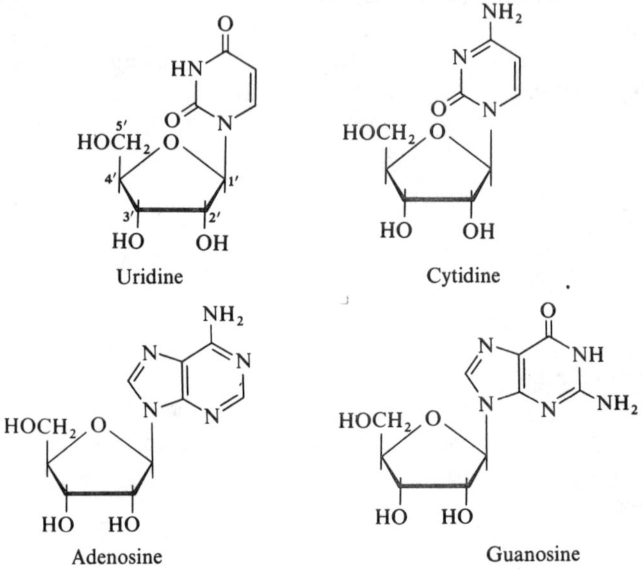

Uridine

Cytidine

Adenosine

Guanosine

SCHEME 9.2. Nucleosides from RNA

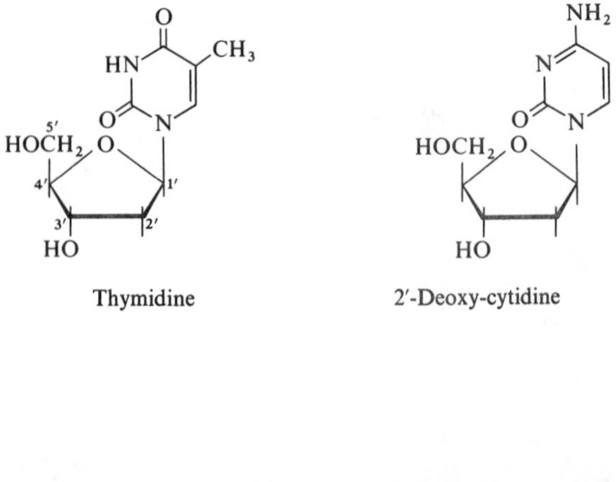

Thymidine

2'-Deoxy-cytidine

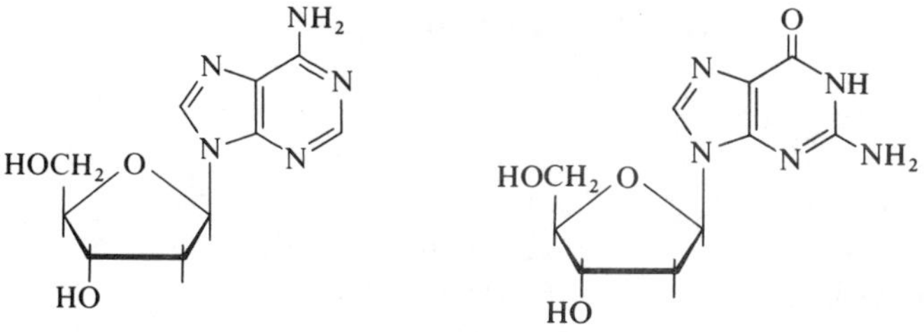

2'-Deoxy-adenosine   2'-Deoxy-guanosine

SCHEME 9.3. Nucleosides from DNA

Amino sugars

Amino-sugars are compounds in which an alcoholic hydroxyl group has been replaced by an amino function. They are named by placing 'x-amino-x-deoxy' in front of the sugar name, where x is the carbon atom bearing the amino group. Amino-sugars occur widely in nature as simple compounds, as components of larger molecules and in polysaccharides. One important class of compounds containing amino-sugars is the carbohydrate antibiotics (see Chapter 12).

The amino-sugar encountered most commonly as a component of natural products is 2-amino-2-deoxy-D-glucose **9.4**, also known as D-glucosamine or chitosamine. Natural monosaccharide deriva-

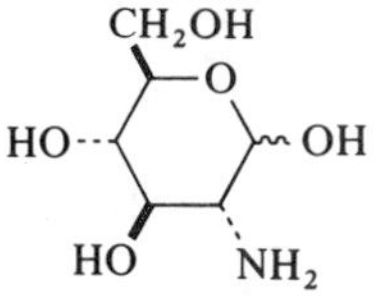

9.4

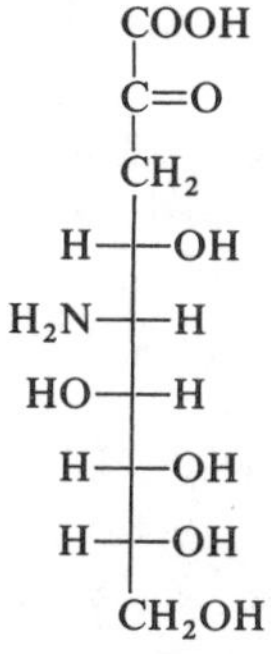

(Open-chain form)

9.5

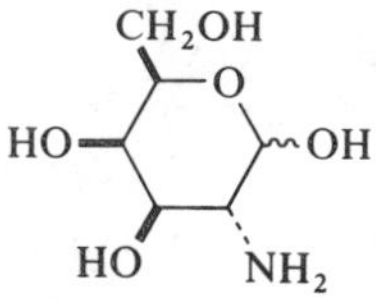

9.6

tives are known in which the amino group is at C-2, C-3, C-4, C-5 or C-6, as well as ones with two amino groups. Compounds bearing

–NHMe or –NMe$_2$ groups have also been isolated. More complex amino-derivatives are also found: particularly important are the sialic acids. The most important of these is neuraminic acid **9.5**.

Because of their natural occurrence many methods have been devised for the synthesis of amino-sugars.

(*a*) Some compounds can be obtained from natural materials. 2-Amino-2-deoxy-D-glucose **9.4** is best prepared from the hydrolysis of chitin, the polysaccharide in the shells of lobsters, crabs, etc. Similarly 2-amino-2-deoxy-D-galactose **9.6** (galactosamine or chondrosamine) can be prepared by acid hydrolysis of chondroitin, a polysaccharide constituent of cartilage and nasal mucilage.

(*b*) A general method for the synthesis of 2-amino-2-deoxy sugars is the addition of hydrogen cyanide to an aldose in the presence of ammonia or an amine, to give an epimeric pair of α-amino-nitriles (E. Fischer 1903, Wolfrom 1946, Kuhn 1955). These products can then be hydrogenated under acidic conditions (see Scheme 9.4).

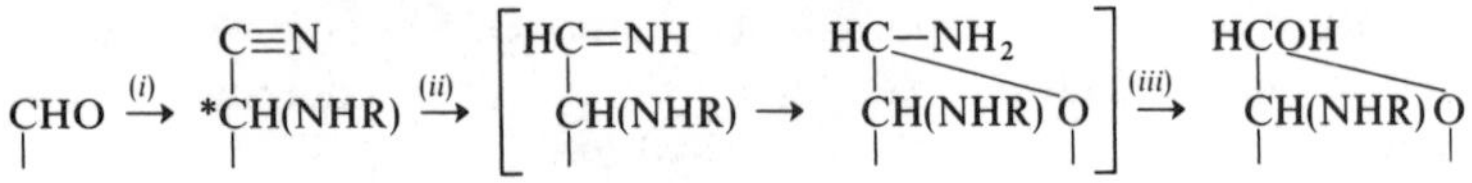

$$
\begin{array}{ccc}
& C\equiv N & \\
| & | & \\
CHO \xrightarrow{(i)} & {}^{*}CH(NHR) & \xrightarrow{(ii)} \\
| & | &
\end{array}
\left[
\begin{array}{cc}
HC=NH & HC-NH_2 \\
| & | \quad | \\
CH(NHR) \rightarrow & CH(NHR)\ O \\
| & |
\end{array}
\right]
\xrightarrow{(iii)}
\begin{array}{c}
HCOH \\
| \quad | \\
CH(NHR)O \\
|
\end{array}
$$

(*i*) RNH$_2$, EtOH, HCN; (*ii*) H$_2$, Pt; (*iii*) H$^{\oplus}$

SCHEME 9.4

If R = Ph or CH$_2$Ph it will be removed during the hydrogenolysis. All possible 2-amino-2-deoxy-hexoses and -pentoses have been synthesized by this method.

(*c*) Perhaps the most general approach is the displacement of a sulphonyloxy group by a nitrogenous nucleophile, usually azide ion, or less commonly, ammonia, an amine, or hydrazine. Such reactions are discussed in detail in Chapter 7. If the nucleophile is azide ion or hydrazine it is readily converted into the required amino function.

(*d*) The attack of the nucleophiles mentioned in (*c*) on sugar oxirans (epoxides) yields products that if not already amino-sugars are readily reduced to them. These reactions follow the general principles described for oxiran ring openings in Chapter 6. The most commonly used nucleophile is azide ion (Guthrie 1963). For example, the 2,3-anhydro-mannoside **9.7** yields methyl 3-azido-4,6-*O*-benzylidene-3-deoxy-α-D-altroside **9.8**.

(*e*) A *trans*-amino-alcohol system, such as that obtained from oxiran opening, can be converted into the corresponding *cis* system

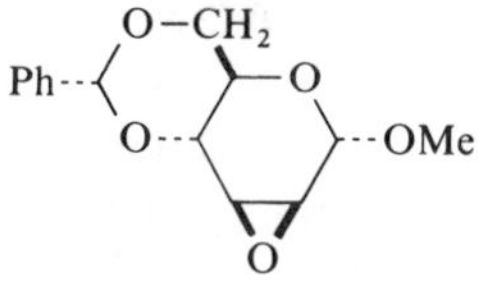

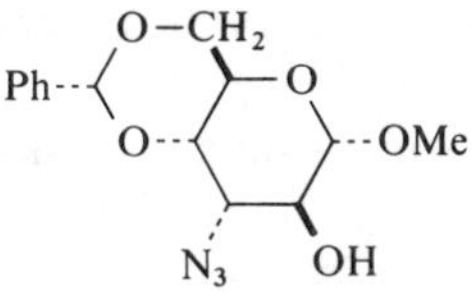

9.7

9.8

by making use of the neighbouring group participation (see p. 45) of an acylamido group. Thus the sequence shown in Scheme 9.5 is

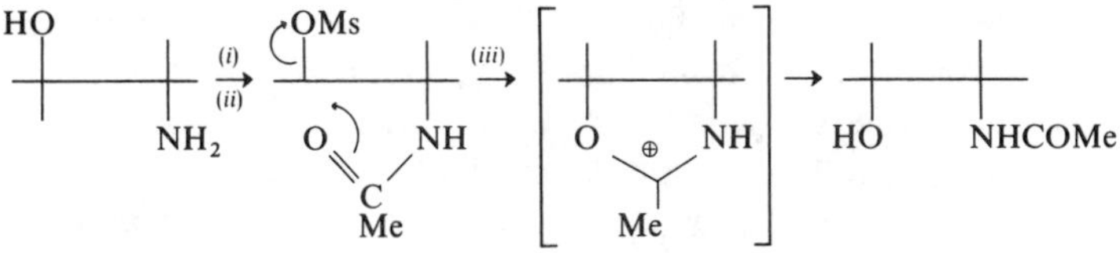

(*i*) Ac$_2$O, MeOH; (*ii*) MsCl, pyridine; (*iii*) KOAc, H$_2$O, MeOCH$_2$CH$_2$OH

SCHEME 9.5

used. For example, the altroside **9.8** can be converted into the alloside compound **9.10** *via* the 3-acetamido-3-deoxy-2-*O*-mesyl-altroside **9.9**.

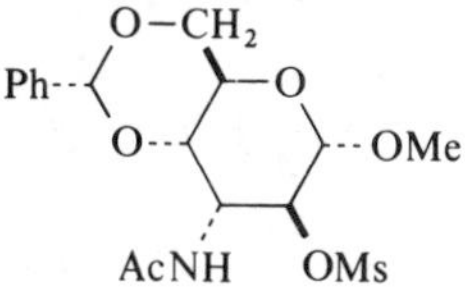

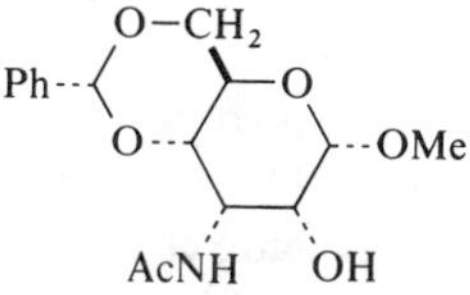

9.9

9.10

(*f*) Periodate-oxidized glycoside derivatives, (see Chapter 8) can be ring-closed using nitromethane in the presence of a base to give cyclic nitro compounds that are then reducible to the corresponding amino-sugar (H. O. L. Fischer 1958). For example, periodate-oxidized methyl α-D-glucopyranoside **9.11** yields the mixture of

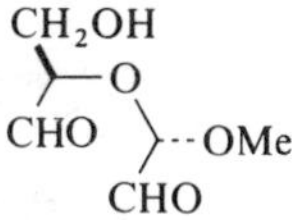

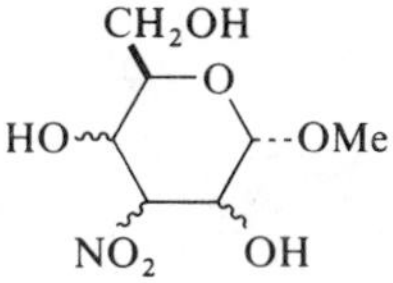

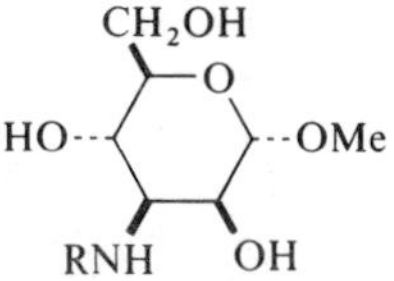

9.11

9.12

9.13 R = H
9.14 R = Ac

products **9.12** from which the mannose derivative **9.13** can be isolated in a worthwhile yield.

Amino-sugars undergo many of the reactions of their parent hydroxyl analogues, for example mutarotation and osazone formation (see below), even if the amino group is on C-2.

Because of the greater reactivity of the amino group when compared to an alcoholic hydroxyl group, selective reactions can be carried out. Treatment of an amino-sugar glycoside with a carboxylic acid anhydride in methanol yields the *N*-acylated product: the hydroxyl groups are unaffected. Thus, methyl 3-amino-3-deoxy-α-D-mannoside **9.13** with methanolic acetic anhydride yields the 3-acetamido-3-deoxy product **9.14**. An alternative route to such products is to fully acetylate the molecule with acetic anhydride in pyridine, followed by de-*O*-acetylation with methanolic sodium methoxide.

Unlike the free sugars, amino-sugars do not yield glycosides on treatment with acid-alcohol mixtures. This is because of the conversion of the C-2-NH_2 group to C-2-NH_3 under these conditions, which provides an effective electrostatic shield, preventing reaction at C-1. If, however, the amino function is first converted into some less basic form, such as an acetamido group, then glycosidation will occur. These phenomena are also manifested in the much greater difficulty in hydrolysing a 2-amino-2-deoxy-glycoside than its C-2-OH or C-2-NHCOR analogues.

Amino-sugars are cleaved by periodate in a similar manner to the corresponding diol (Chapter 8).

A sub-class of amino-sugars are 4-amino and 5-amino-derivatives in which the amino function may form part of the ring system in place of the ring oxygen atom. These are too specialized to be dealt with in detail here, but it is of interest to note that 5-amino-5-deoxy-D-glucose (nojirimycin) **9.15** occurs naturally and has antibiotic properties.

9.15

Derivatives of phenylhydrazine

The reactions of phenylhydrazine with monosaccharides were first studied by E. Fischer (1884) for structural correlation between different monosaccharides, and for characterizing individual sugars. Although the reaction has been known for a long period, it is only recently that its mechanism has been begun to be understood. With one equivalent of phenylhydrazine a sugar behaves as a normal aldehyde or ketone and forms a phenylhydrazone. However, with excess of reagent (three equivalents required), osazones are formed: for example D-glucose gives the phenylosazone **9.16**; ammonia and aniline are formed as by-products. It should be noted that in forming an osazone the asymmetry at C-2 has been destroyed. Thus, D-glucose and D-mannose both yield the same phenylosazone **9.16**. Further, the same compound **9.16** also results from the treatment of D-fructose with phenylhydrazine. The use of the reaction for structural determination can be easily seen. The osazones appear

9.16   9.17   9.18

from the X-ray and n.m.r. data to have a chelated ring structure such as **9.17**, which also accounts for chemical properties such as formation of only an *N*-methyl rather than a di-*N*-methyl derivative. Formation of a chelated ring structure is one reason why the osazone reaction does not continue all down the chain. In line with this, is the recent observation (Chapman, 1967) that aldoses when treated with large excess of hydrazines such as 1-methylphenylhydrazine yield alkazones, such as **9.18**.

Removal of the phenylhydrazine residues from an osazone with excess of benzaldehyde in acetic acid yields a 1,2-dicarbonyl sugar, called an *osone*. Thus, D-glucose phenylosazone **9.16** when treated in this way yields glucosone **9.19**.

Mild oxidation of osazones, with reagents such as cupric sulphate, yields *osotriazoles*, used occasionally to characterize sugars; for example, D-glucose phenylosazone **9.16** gives the osotriazole **9.20**.

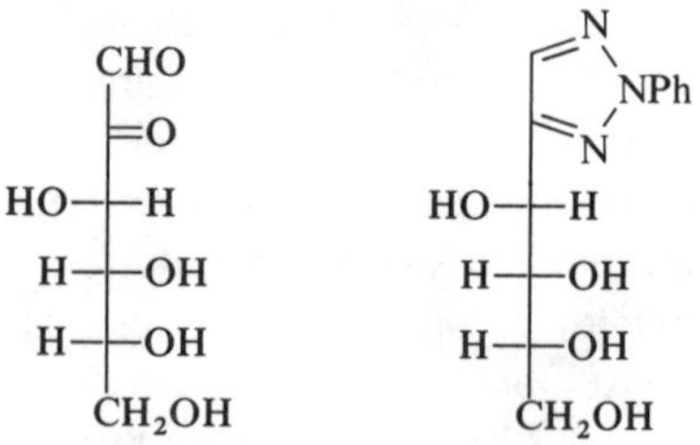

9.19   9.20

Other nitrogen-containing monosaccharide derivatives

There are many other types of monosaccharide derivatives bearing a sugar carbon–nitrogen bond. Nitro-sugars, usually prepared as described on p. 59 are useful precursors for amino-sugars.

Another group are the sugar epimines or aziridines **9.21** (Guthrie, Hough, Richardson, 1965) which are nitrogen analogues of oxirans.

9.21   9.22

They are generally synthesized from the system **9.22** by some reductive process. Ring-opening reactions of the three-membered ring leads to amino-sugar derivatives.

10 Other monosaccharide derivatives

IN the chapters so far, the most important monosaccharide derivatives have been discussed. Other derivatives are known, many of which have been studied in some detail. However, in an introductory text they must be treated briefly, and some of the various types and their principal points of interest are discussed below.

Unsaturated sugars

In recent years several natural products have been isolated that have contained unsaturated sugar components. The compounds found have had the double bond in the 2,3- or the 4,5-position in a pyranose ring, or in the 4,5-position in a furanose ring. Thus both endocyclic and exocyclic systems occur in nature. Examples are Blasticidin S **10.1** angustmycin A **10.2** and part of the sisomicin molecule **10.3**.

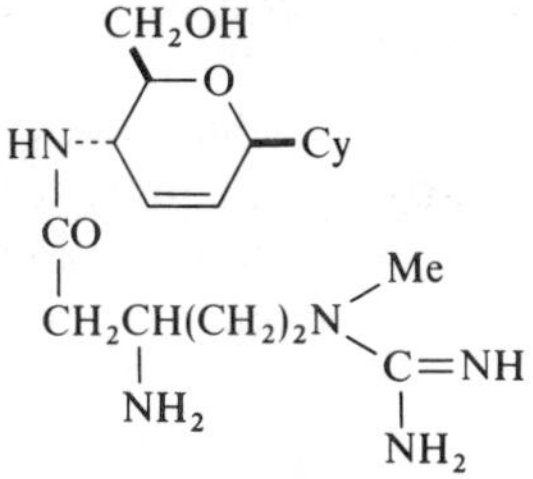

Cy = cytosine

10.1

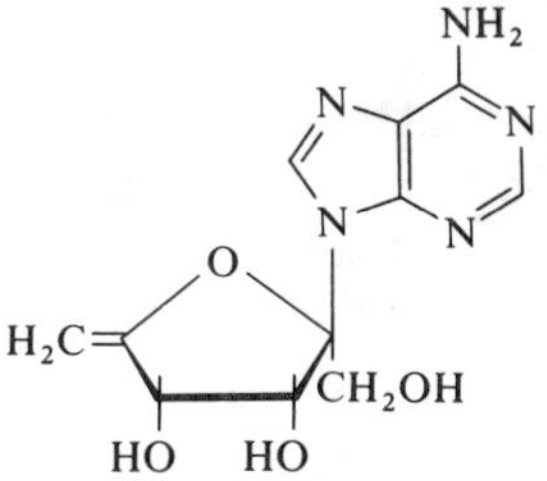

10.2

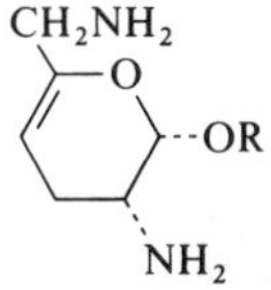

10.3

Like simple alkenes, unsaturated sugars have been found to be valuable intermediate in synthesis. Compounds have been made in which the carbon–carbon double bond is either exo- or endocyclic to ring systems of both sizes and in acyclic compounds. Triple bonds have also been introduced into acyclic systems. We shall only consider double bond systems and shall do so in two parts: firstly, the so-called glycals, that have a 1,2-double bond, and secondly, those compounds with the double bond elsewhere in the molecule.

Glycals

Acetobromaldoses on reaction with zinc and acetic acid give acetylated glycals (E. Fischer 1914, Helferich 1954). Thus, acetobromoglucose **10.4** gives tri-*O*-acetyl-D-glucal **10.5** which can be de-acetylated in the usual way.

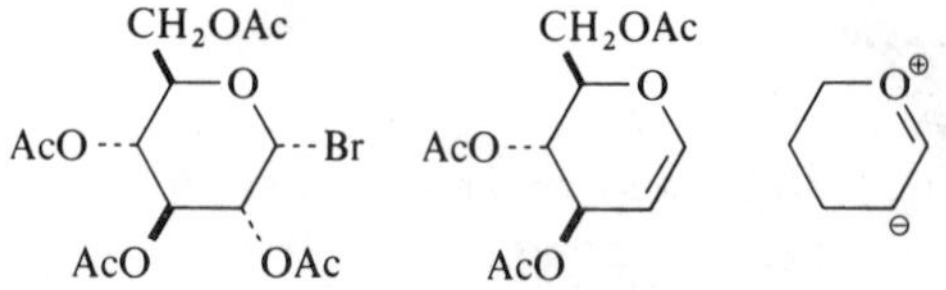

10.4   **10.5**   **10.6**

Glycals, which are vinyl ethers, readily undergo addition reactions. The direction of addition is controlled by the contribution of the mesomeric form **10.6**, so that electrophilic attack occurs at C-2. In this way acid-catalysed addition of water or alcohols to glycals gives 2-deoxy-aldose derivatives. Reaction of alcohols in the presence of boron trifluoride etherate with tri-*O*-acetyl-D-glycals causes attack at C-1 with allylic migration of the double bond and expulsion of the C-3-OAc group (Scheme 10.1); the α-anomer predominates in

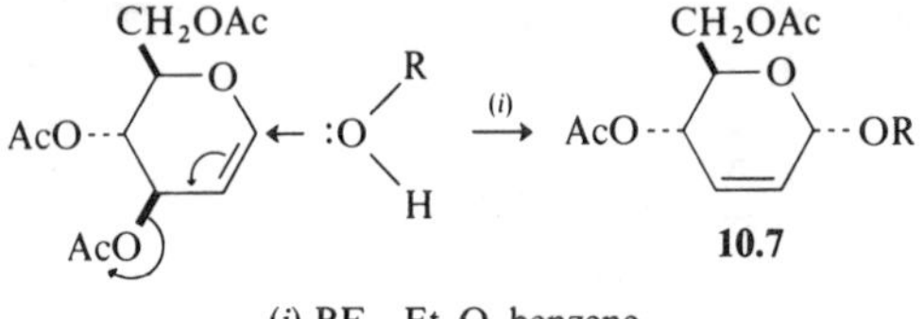

10.7

(*i*) BF$_3$, Et$_2$O, benzene

SCHEME 10.1

the product (Ferrier, 1968). Compounds such as **10.7** are useful intermediates for further conversions.

Another important addition reaction of glycals is the reaction with nitrosyl chloride to give dimeric products, useful for the synthesis of α-glycosides (see Chapter 4).

Hydroxylation of glycals with perbenzoic acid gives 1,2-dihydroxyl compounds *via* an intermediate oxiran (epoxide). As in simpler alicyclic systems the neighbouring hydroxyl group on C-3 directs the formation of the oxiran by hydrogen-bonding, to its side of the molecule; the oxiran ring is then opened predominantly at the more reactive C-1 position. Thus, D-glucal gives D-mannose and a trace of D-glucose, (Scheme 10.2), and similarly D-galactal gives

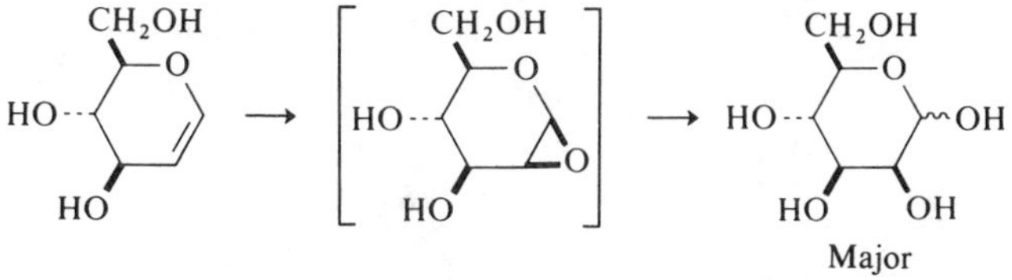

SCHEME 10.2

mostly D-talose, with some D-galactose. Blocking of the hydroxyl groups, for example by acetylation, removes the hydrogen-bonding directing effect and so 3,4,6-tri-*O*-acetyl-D-glucal **10.4**, for example, gives mainly 3,4,6-tri-*O*-acetyl-D-glucose. It is assumed here that the C-3-OAc shields the β-face of the molecule and so the inter-mediate epoxide has the configuration **10.8**.

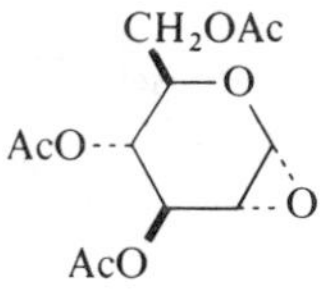

10.8

Other olefinic sugars

The most-studied compounds in this class are the hex-2-enopyran-oses with the double bond between C-2 and C-3, though examples of all types with endocyclic double bonds have been described. Some of the general methods for synthesis are shown in Scheme 10.3. The Tipson–Cohen reaction (1965) of vicinal disulphonates with zinc and sodium iodide in DMF has found wide use, and either *cis* or *trans* precursors can be used. Probably the best method from *cis*-diols is the Corey–Winter method via the thionocarbonate (Corey, 1963, Haines, 1965, Horton, 1966). Aldo-2-enopyranosides can also be prepared from glycals (see above).

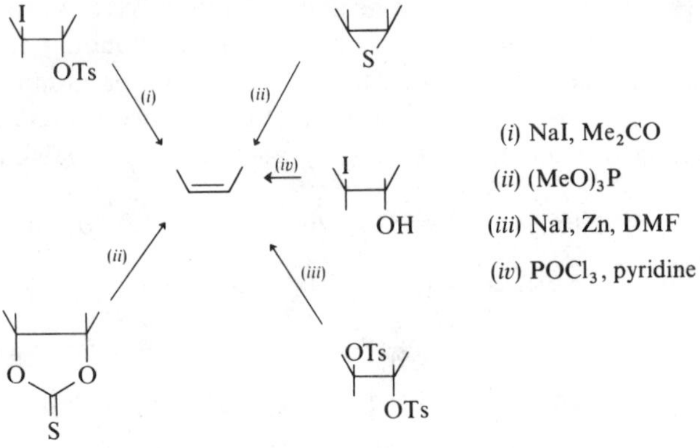

(i) NaI, Me$_2$CO

(ii) (MeO)$_3$P

(iii) NaI, Zn, DMF

(iv) POCl$_3$, pyridine

SCHEME 10.3

Exocyclic aldohex-5-enopyranoses are available from treatment of suitably blocked 6-iodo-6-deoxy derivatives with silver fluoride in pyridine (Scheme 10.4) (Helferich 1928).

Hydrogenation of olefinic sugars provides a useful route to vicinal dideoxy sugars.

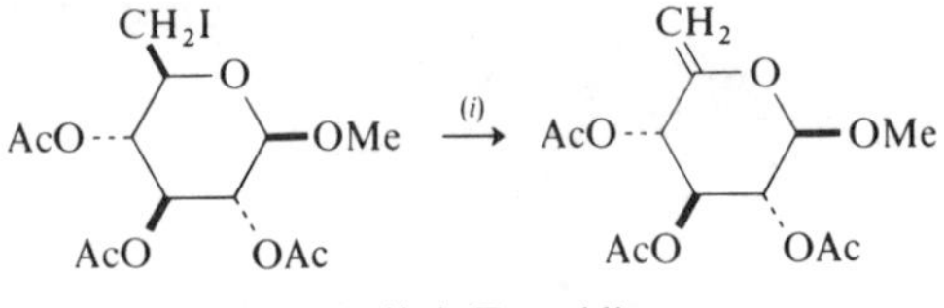

(i) AgF, pyridine

SCHEME 10.4

Deoxy-Sugars

Deoxy-sugars are those compounds in which one or more hydroxyl groups have been replaced by hydrogen. Mono-, di-, and tri-deoxy-sugars occur naturally. The 6-deoxy-aldohexoses occur widely, for example 6-deoxy-L-mannose (L-rhamnose), and 6-deoxy-L-galactose (L-fucose). The synthesis of the 6-deoxy function is most easily achieved by lithium aluminium hydride reduction of a suitably blocked 6-bromo-, 6-iodo, or 6-sulphonyloxy compound. An example is the formation of 6-deoxy-1,2:3,4-di-O-isopropylidene-α-D-galactose **10.10** from the parent alcohol **10.9**, *via* the tosylate (Scheme 10.5).

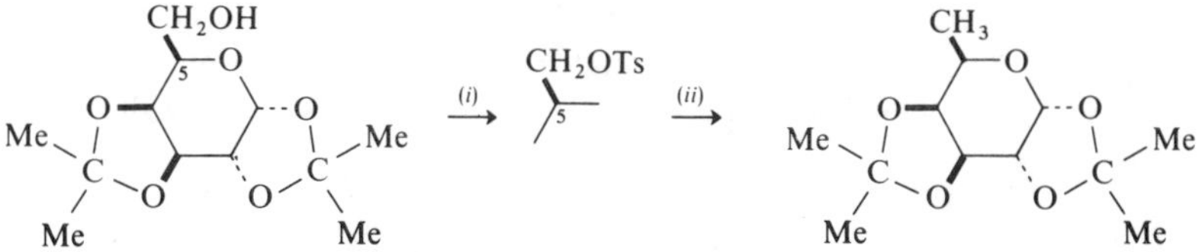

10.9   (*i*) TsCl, pyridine; (*ii*) LiAH₄, Et₂O   **10.10**

SCHEME 10.5

Deoxy sugars may be prepared by reduction of oxirans with lithium aluminium hydride. For example, the 2,3-anhydro-alloside **10.11** gives predominantly the 2-deoxy-compound **10.12** (diaxial opening) (Prins, 1948) (see also p. 37).

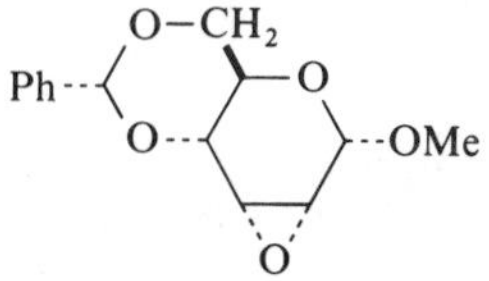

10.11 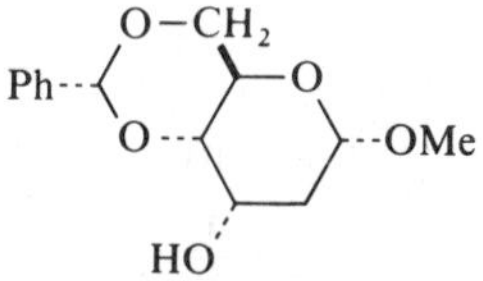   **10.12** 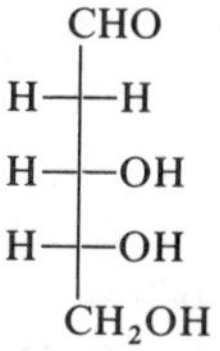   **10.13**

The most important 2-deoxy-aldose is 2-deoxy-D-*erythro*-pentose,† or 2-deoxy-D-ribose as it is more commonly, but incorrectly, called; its structure is shown in the open chain form in **10.13**. 2-Deoxy-D-*erythro*-pentose occurs widely as the sugar component of deoxyribonucleic acids (DNA) in which it exists in the furanose form (see p. 56). 2-Deoxy-aldose derivatives can be prepared from glycals as described above. Glycosides of 2-deoxy-aldoses are very rapidly hydrolysed by aqueous acid to the free sugar (see p. 23).

Several types of dideoxy-sugars are found naturally. 3,6-Dideoxy-hexoses are components of the polysaccharides of *Salmonella* species. 2,6-Dideoxy-hexoses often with a branch at C-3 occur as components of some antibiotics, as do less commonly 3,4-dideoxy-aldohexose derivatives.

Branched-chain sugars

Such compounds have a carbon branch somewhere along the sugar chain. Branching can be of type **10.14** in which the OH group

† See footnote on p. 50.

is replaced, or **10.15** in which the H is replaced. R is generally CH_3, CH_2OH, or CHO in naturally-occurring examples, which include for example D-apiose (3-*C*-hydroxymethyl-D-*glycero*-tetrose)

$$H\!-\!\!|\!-\!R \qquad HO\!-\!\!|\!-\!R$$

10.14 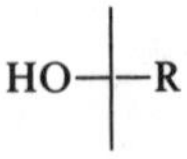   **10.15**

10.16 from parsley, and streptose (5-deoxy-3-*C*-formyl-L-lyxose) **10.17**, a component of the antibiotic streptomycin (see p. 73).

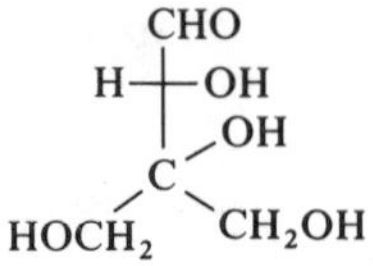

10.16 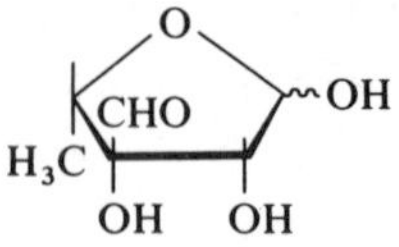   **10.17**

One of the general synthetic approaches to branch-chain sugars is *via* carbonyl compounds, either by attack with Grignard reagents or similar organometallic compounds, or by treatment with diazomethane to give spiroepoxides that can be opened to give branched compounds (Scheme 10.6).

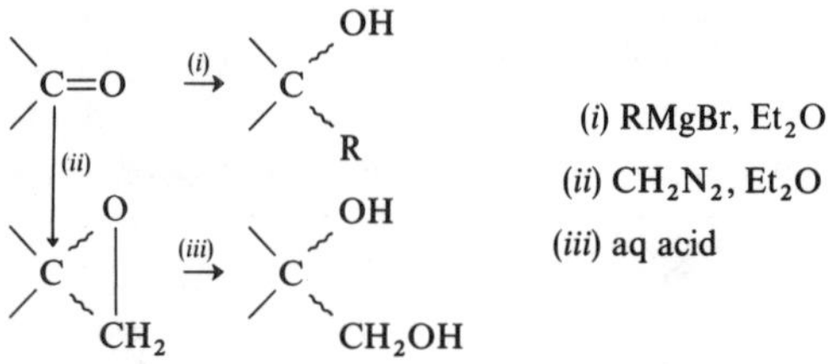

(*i*) $RMgBr$, Et_2O

(*ii*) CH_2N_2, Et_2O

(*iii*) aq acid

SCHEME 10.6

Sugar acids

Some of these have been dealt with in Chapter 8. An important member of this class is L-ascorbic acid (vitamin C), first isolated as a crystalline substance by Szent Gyorgyi in 1928, and now known to have structure **10.18**. The majority of ascorbic acid consumed is synthetic, prepared by a multistep synthesis from D-glucose (Reichstein, 1934). A recent route (Theander, 1971) starting from 1,2-*O*-

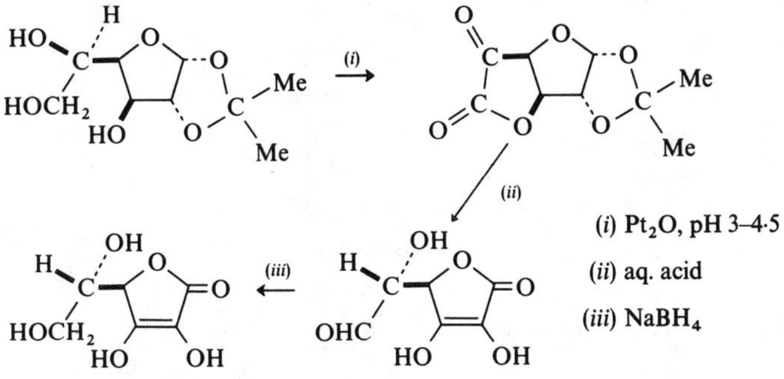

10.18

SCHEME 10.7

isopropylidene-D-glucose makes the synthesis even simpler (Scheme 10.7).

11 Alditols and cyclitols

THESE two classes of compounds do not strictly fall within the definition of carbohydrates as given in Chapter 1. Nevertheless their chemistry is so similar to that of monosaccharides that it is customary to consider them alongside carbohydrates.

Alditols

Alditols are polyhydroxyalkanes, generally prepared by reduction of aldoses or ketoses. Methods used have been sodium amalgam, electrolytic reduction, or catalytic high-pressure hydrogenation. The best laboratory method is reduction with sodium borohydride in aqueous solution. In this way D-glucose yields D-glucitol (D-sorbitol) **11.1**, and D-galactose gives D-galactitol (dulcitol) **11.2**. Reduction of a ketose yields a mixture of two alditols, because of

```
        CH2OH              CH2OH              CH2OH
   H ──┼── OH         H ──┼── OH        HO ──┼── H
  HO ──┼── H         HO ──┼── H        HO ──┼── H
   H ──┼── OH        HO ──┼── H         H ──┼── OH
   H ──┼── OH         H ──┼── OH        H ──┼── OH
        CH2OH              CH2OH              CH2OH

        11.1               11.2               11.3
```

the creation of a new chiral centre at C-2; thus, D-fructose gives a mixture of D-glucitol **11.1** and D-mannitol **11.3** (C-2-epimers).

Fewer stereoisomers exist for each series than for the corresponding aldoses, because the end groups in the molecule are identical. This also means that *meso* (internally compensated) forms are found

```
     CH2OH           CH2OH            CH2OH
  H ─┼─ OH         H ─┼─ OH        HO ─┼─ H          CH2OH
  H ─┼─ OH        HO ─┼─ H         H ─┼─ OH       H ─┼─ OH
  H ─┼─ OH         H ─┼─ OH        H ─┼─ OH          CH2OH
     CH2OH           CH2OH            CH2OH

     11.4            11.5             11.6             11.7
```

(these are not optically active). There are only four pentitols, compared with eight stereoisomers for the aldopentoses: ribitol (*meso*) **11.4** xylitol (*meso*) **11.5**, D-arabinitol **11.6** and its enantiomorph. There are ten hexitols compared with sixteen aldohexoses.

The simplest alditol is generally considered to be glycerol **11.7**, which occurs widely in long-chain fatty esters of oils and fats. Ribitol, occurs in teichoic acids, components of some bacterial cell walls (see p. 95). D-Mannitol occurs widely in seaweeds and D-glucitol in the berries of the mountain ash.

As would be expected, alditols undergo many typical reactions of the alcoholic hydroxyl groups of monosaccharides, such as esterification, acetal formation, and oxidation. An interesting aspect of recent studies on alditols and their derivatives has been the work on the investigation of their detailed conformation (see Chapter 3).

Cyclitols

The most studied members are the inositols or cyclohexanehexols. Nine stereoisomeric structures exist as shown in Scheme 11.1. They are named by placing the prefix shown before 'inositol': for

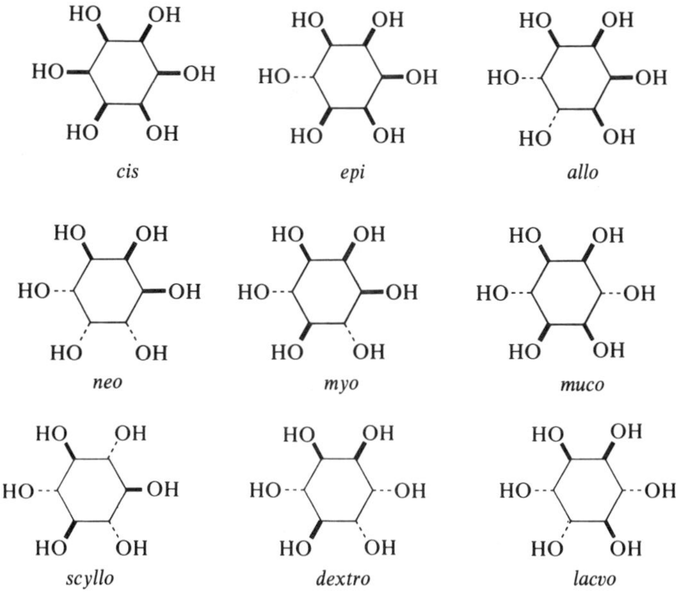

SCHEME 11.1

example *cis*-inositol. Each exists preferentially in the conformation with the most equatorial hydroxyl groups; all but two of the inositols are optically inactive.

Inositols undergo many of the reactions of monosaccharides. Isopropylidene acetals can be prepared, pairs of vicinal *ax,eq* hydroxyl groups generally being involved. It is possible to form such an acetal ring bridging *eq,eq*-hydroxyl groups, though this system is more strained (Angyal, 1952). The various types of esters discussed in Chapter 7 can be formed and show similar reactions; for example displacement of sulphonates, either directly or with neighbouring group participation, and migration of carboxylic esters.

Other inositol derivatives are the deoxy-inositols (quercitols) such as *cis*-quercitol **11.8** and the inososes, such as *epi*-inosose **11.9**. An important class of derivatives is the amino-inositols, particularly

11.8   **11.9**   **11.10**   **11.11**

streptamine **11.10** and 2-deoxy-streptamine **11.11** which are found as components of antibiotics (see Chapter 12).

12 Carbohydrate antibiotics

A NEW area of carbohydrate chemistry was opened up when it was found that many naturally-occurring carbohydrate compounds possessed antibiotic properties. The first such substance to have real impact on chemotherapy was streptomycin **12.1**, first isolated in

12.1

1944. It can be seen to be made up from three units: 2-methylamino-2-deoxy-L-glucose; a branched-chain sugar, streptose; and an amino-cyclitol, streptamine. Streptomycin, in combination with other drugs is very effective for the treatment of TB.

In the last fifteen or so years the number of naturally-occurring carbohydrate compounds with antibiotic properties has increased tremendously. In this text little more can be done than review the area briefly. In Chapter 16 examples are given of the detailed methods used for structural and synthetic work for such systems.

The carbohydrate-containing antibiotic substances fall into several broad classes. Some of these are compounds that are largely or completely carbohydrate, whilst others have the carbohydrate as only a small part of the molecule. Quite often the monosaccharides involved are amino-sugars (see p. 57).

Nucleoside antibiotics

Nature seems only to be able to tolerate small changes from the 'normal' nucleosides. Thus, with very few exceptions, the antibiotics of this class either have a 'normal' base with a sugar other than D-ribofuranose, or an unnatural base coupled to β-D-ribofuranose. The latter type are often '*C*-glycoside' analogues of natural nucleosides. Examples of this general class are cordycepin **12.2**, and formycin **12.3** (see also formulae **10.1** and **10.2** p. 63).

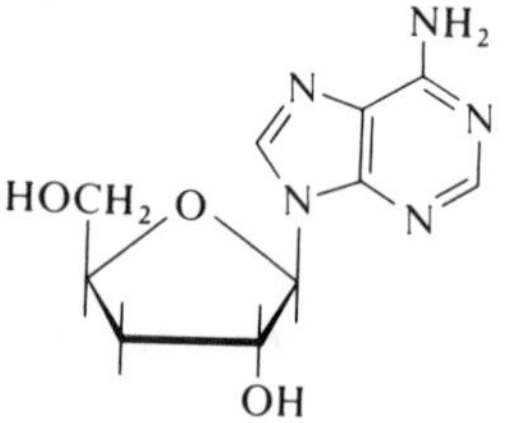

12.2 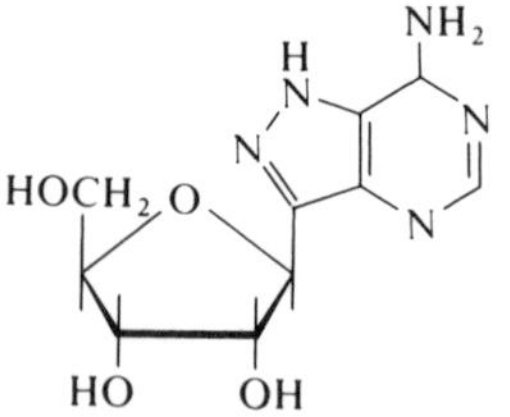   **12.3**

Antibiotics containing aromatic groups

These contain one or two sugars attached to an aromatic system, generally polycyclic in character. An example is daunamycin **12.4**, which contains the amino-sugar daunosamine.

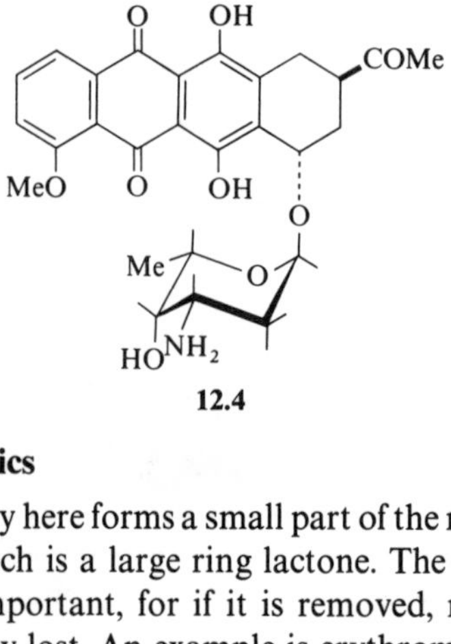

12.4

Macrolide antibiotics

The sugar moiety here forms a small part of the molecule, the main component of which is a large ring lactone. The role of the carbohydrate part is important, for if it is removed, most or all of the activity is generally lost. An example is erythromycin **12.5**.

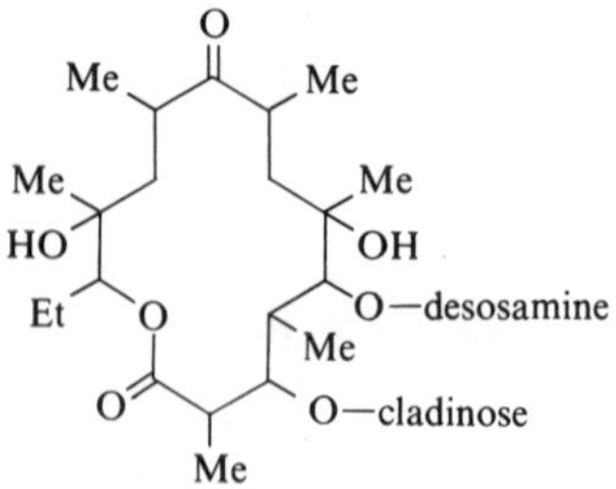

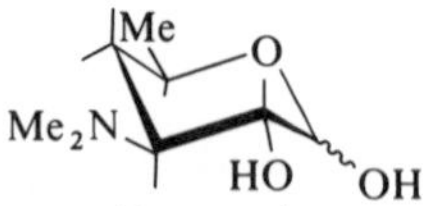

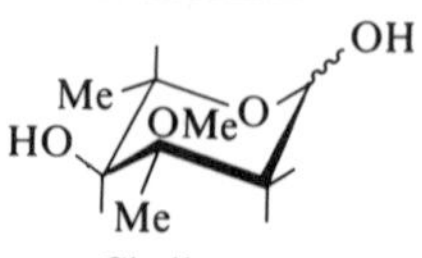

12.5

Desosamine

Cladinose

Amino-glycoside antibiotics

This is the class that is perhaps of most interest to carbohydrate chemists. The general structural features of the substances are an amino-cyclitol to which are linked one or more amino-sugars and sometimes other sugars. Streptomycin **12.1** falls into this group, though the two main classes are typified by neomycin B **12.6** and kanamycin A **12.7**; note that the glycosidic linkages in this class are

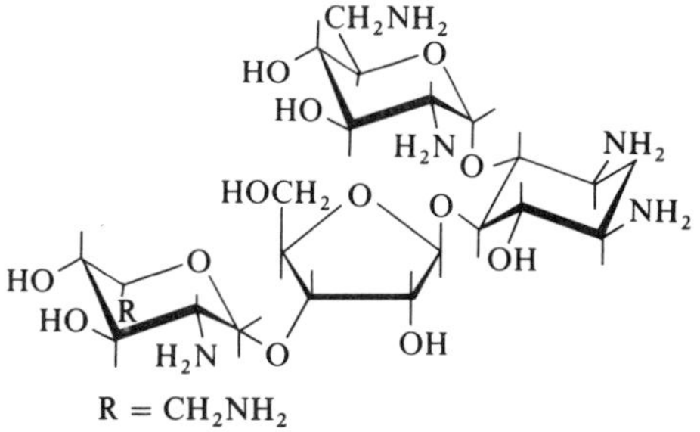

$R = CH_2NH_2$

12.6

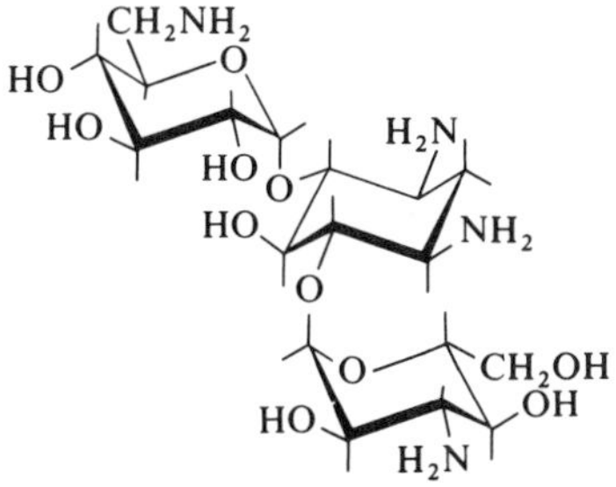

12.7

generally α. The chemotherapeutic interest in such compounds is that they are active against bacteria that are resistant to other antibiotics such as the various penicillins. Other members of this class used clinically are the gentamicin C† complex **12.8** and tobramycin **12.9**. Quite subtle changes in structure can have quite marked effects on the activity. Both gentamicin C **12.8** and tobramycin **12.9** are active against strains resistant to kanamycin A **12.7**. This is because the first two have no hydroxyl group at C-3 in the

† This is not a typographical error: substances elaborated by *Streptomyces* species are given the ending 'mycin'; those from other sources, the suffix 'micin'.

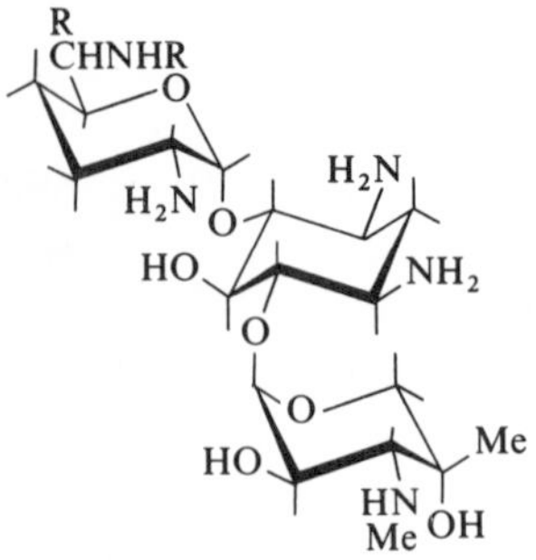

12.8

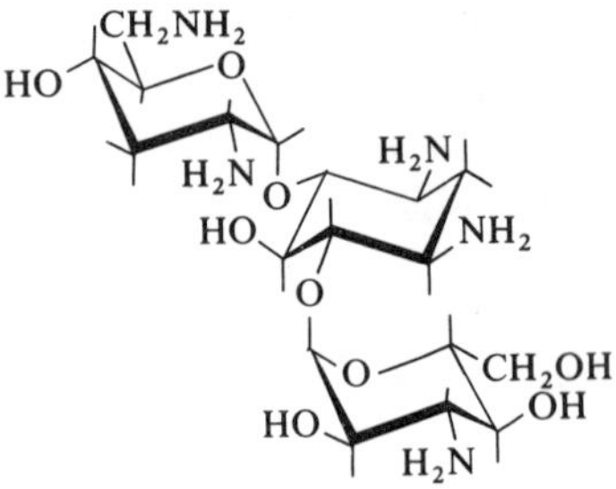

12.9

diamino-sugar unit. In kanamycin A this can be enzymically phosphorylated to give an inactive substance.

Miscellaneous antibiotics

Some are quite simple compounds, such as nojirimycin already mentioned on p. 60 and streptozotocin **12.10**, a simple derivative of

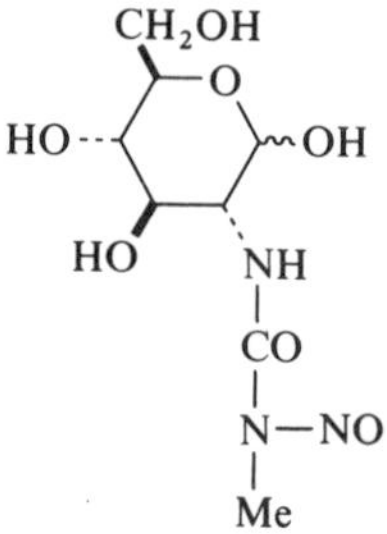

12.10

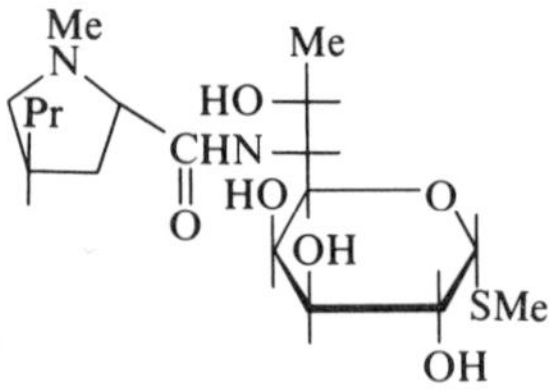

12.11

2-amino-2-deoxy-D-glucose. A chemotherapeutically important antibiotic with novel structural features is lincomycin **12.11**; note that it is a thioglycoside of a 6-amino-6-deoxy-octose.

13 Oligosaccharides

OLIGOSACCHARIDES are the group of carbohydrates which are intermediate in size between the simple monosaccharides and the polymeric polysaccharides. The name is generally applied to those substances with two to ten monosaccharide units ('oligo': few) interlinked by glycosidic bonds. Because of these glycosidic linkages, oligosaccharides are readily hydrolysed by aqueous acid to their constituent monosaccharides. Many important natural products are oligosaccharides.

Disaccharides

Only two, sucrose and lactose, are found abundantly in nature. The former, the common table sugar, or just 'sugar', is obtained commercially from sugar beet or cane. Lactose ('milk sugar') occurs in mammalian milk. Maltose, found free in nature to a limited extent (for example in soya beans), is normally prepared by enzymic hydrolysis of starch. Cellobiose is obtained by partial hydrolysis of cellulose.

The disaccharides are glycosides (see Chapter 4) in which the aglycone is another monosaccharide. When the inter-sugar link is between the reducing (anomeric) centre of one unit and an alcoholic hydroxyl group of the other, as in **13.1**, the substance is said to be a *reducing* disaccharide, as it has a potential aldehyde group in the molecule, and will react with reagents such as Fehling's solution. If the glycosidic link is between the two anomeric centres, as in **13.2**, then the substance is said to be a *non-reducing* disaccharide. Both types are found naturally.

Reducing disaccharides are named as substituted monosaccharides. Consider **13.1**: the right-hand monosaccharide unit (*xylo*) is

13.1   **13.2**

considered as the fundamental unit for naming, as it has its reducing centre free, and so **13.1** is 4-*O*-(α-D-glucopyranosyl)-D-xylopyranose.

For a non-reducing disaccharide either monosaccharide unit can be considered as the basic unit, and the other as the aglycone. Thus **13.2** may be named either as β-D-xylopyranosyl α-D-glucopyranoside, or as α-D-glucopyranosyl β-xylopyranoside. This nomenclature makes it clear that the anomeric centre of each monosaccharide is involved and so no numbers are included in the names.

A shortened notation is sometimes employed that is especially useful for higher oligosaccharides (and for polysaccharides). This uses the first three letters of the monosaccharide names, except for glucose (G); and *p* and *f*, for pyranose and furanose respectively, together with appropriate numbers to denote the linkages. For example, the disaccharide **13.1** would be described as α-D-G*p* 1–4 D-Xyl*p*.

The information that must be obtained for the determination of the structure of a disaccharide is: (1) identity of the monosaccharide residues; (2) sequence and ring-size of the residues; (3) the inter-unit linkage positions; and (4) whether the glycosidic linkage is α or β.

The normal procedure for the structural investigation of a disaccharide is therefore as follows: (*a*) Determination whether the disaccharide is reducing or non-reducing, by testing with reagents such as Fehling's solution. (*b*) Acid hydrolysis to the parent monosaccharide(s), which are identified. This can be done by paper chromatography, or by gas–liquid chromatography after reduction with sodium borohydride and acetylation. (*c*) Determination of the anomeric linkage. This can be done by investigating the coupling constant of the anomeric protons in the n.m.r. spectrum (see Chapter 15), by enzymic studies using, for example, maltase which hydrolyses α-linkages or emulsin (β-linkages), or by consideration of the specific rotation. (*d*) Complete methylation followed by hydrolysis with aqueous acid gives two partially methylated monosaccharides. These are separated and characterized (often after reduction to the alditols) by gas–liquid chromatography. This generally permits determination of ring-size and point of linkage. Periodate or lead tetra-acetate oxidation studies often provide an additional tool for investigating such features.

An example of the structural study of a disaccharide is given in Chapter 16.

Cellobiose is easily obtained as its octa-acetate by acetolysis of cellulose. In summary, its structure was elucidated as follows: it was reducing, hydrolysis by aqueous acid gave only D-glucose, and it had a β-link (hydrolysed by emulsin and not maltase). Degradation of its octa-*O*-methyl ether and of methyl octa-*O*-methyl-cellobionic acid showed the presence of a 1–4 linkage as in **13.3**. Thus cellobiose is 4-*O*-(β-D-glucopyranosyl)-D-glucopyranose. This assignment has been confirmed by synthesis.

13.3

Sucrose the most common carbohydrate, has the world's largest production for a single pure organic substance. Its structure was deduced by the general methods outlined above as **13.4**, namely

13.4

α-D-glucopyranosyl β-D-fructofuranoside. Acid hydrolysis of sucrose ([α]$_D$ + 66°) gives an equimolar mixture of D-glucose ([α]$_D$ + 52°) and D-fructose ([α]$_D$ − 92°), which, of course has a net negative rotation. For this reason, this hydrolysis is sometimes called the inversion of sucrose and the D-glucose, D-fructose mixture is known as 'invert sugar'.

The chemistry of sucrose has not been studied as much as would be expected for such a readily available and cheap natural material. Its synthesis has only been achieved in low yield; reaction of 3,4,6-tri-*O*-acetyl-1,2-anhydro-α-D-glucopyranose **13.5** with 1,2,3,6-tetra-*O*-acetyl-D-fructofuranose **13.6** in a sealed tube at 100° and subsequent acetylation gave a 5 per cent yield of sucrose octa-acetate (Scheme 13.1) (Lemieux, 1953).

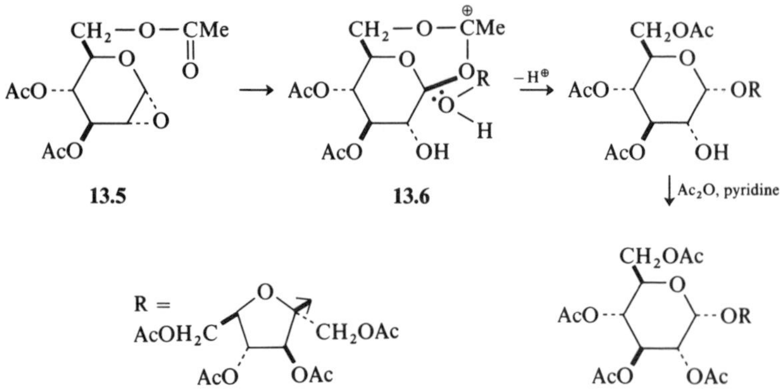

SCHEME 13.1

Other disaccharides

The structures of some other common disaccharides are summarized in Table 13.1. Their structures were established by the general procedures described above.

TABLE 13.1

Some common disaccharides

Disaccharide	Source	Structure
Trehalose	Fungi, yeasts	α-D-glycopyranosly-α-D-glucopyranose
Sophorose	Hydrolysis of a glycoside from *Sophora japonica*	2-*O*-(β-D-glycopyranosyl)-D-glucopyranose
Turanose	Partial hydrolysis of melezitose	3-*O*-(α-D-glucopyranosyl)-D-fructose
Laminaribiose	Partial hydrolysis of laminarin from seaweeds	3-*O*-(β-D-glucopyranosyl)-D-glucopyranose
Maltose	Enzymic hydrolysis of starch	4-*O*-(α-D-glucopyranosyl)-D-glucopyranose
Maltose	Enzymic hydrolysis of starch	4-*O*-(α-D-glucopyranosyl)-D-glucopyranose
Lactose	Whey	4-*O*-(β-D-galactopyranosyl)-D-glucopyranose
Isomaltose	Enzymic hydrolysis of amylopectin	6-*O*-(α-D-glucopyranosyl)-D-glucopyranose
Gentiobiose	Partial hydrolysis of gentianose	6-*O*-(β-D-glucopyranosyl)-D-glucopyranose
Melibiose	Partial hydrolysis of raffinose	6-*O*-(α-D-galactopyranosyl)-D-glucopyranose

Disaccharide synthesis

This is essentially glycoside synthesis but with a complex aglycone. The general methods described in Chapter 4 have been used, in particular the Koenigs–Knorr synthesis. For example gentiobiose (β1–6) was synthesized from 2,3,4,6-tetra-O-acetyl-α-D-glucopyranosyl bromide and 1,2,3,4-tetra-O-acetyl-D-glucopyranose (Reynolds, 1938). Use of a non-participating group, as in 3,4,6-tri-O-acetyl-2-O-nitro-β-glucopyranosyl chloride **13.7**, with the same aglycone gave isomaltose octa-acetate (α1–6 linkage) (Wolfrom, 1961).

13.7

Trisaccharides and higher oligosaccharides

The structural study of these molecules follows the same principles as those described above for disaccharides. Their synthesis presents more problems than does the synthesis of disaccharides. For oligomers of a simple sugar, partial hydrolysis of a polysaccharide followed by chromatography is often the only method. Partial acetolysis of cellulose gives cellobiose, cellotriose **13.8**, etc., up to celloheptaose **13.9** as their peracetates.

13.8 $n = 1$
13.9 $n = 5$

Raffinose is the most abundant trisaccharide found in nature. It occurs in the mother liquors from sugar-beet crystallization and as a crystalline exudate on certain eucalyptus trees. The constitution of this trisaccharide was elucidated by hydrolysis experiments. Dilute acid gave equimolecular proportions of D-glucose, D-galactose, and D-fructose. Treatment with invertase afforded D-fructose and melibiose, whereas α-galactosidase, an enzyme which specifically breaks the α-galactoside link, led to D-galactose and sucrose. Thus, the raffinose molecule is made up of D-glucose and D-galactose linked as in melibiose, with the D-fructose linked to D-glucose as in sucrose. This has been confirmed by methylation studies. Raffinose is thus *O*-α-D-galactopyranosyl-(1–6)-*O*-α-D-glucopyranosyl-(1–2)-β-fructofuranoside (or α-D-Gal*p* 1–6 α-D-G*p* 1–2 β-D-Fru*f*) **13.10**.

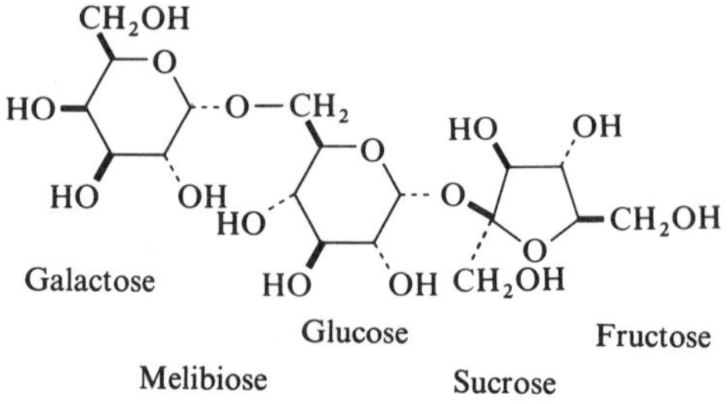

13.10

Gentianose, another non-reducing trisaccharide, obtained from gentian roots, had its structure determined by the methods used for raffinose. It was shown to be *O*-β-D-glucopyranosyl-(1–6)-*O*-α-D-glucopyranosyl-(1–2) β-D-fructofuranoside (or β-D-G*p* 1–6 α-D-G*p* 1–2 β-D-Fru*f*) **13.11**.

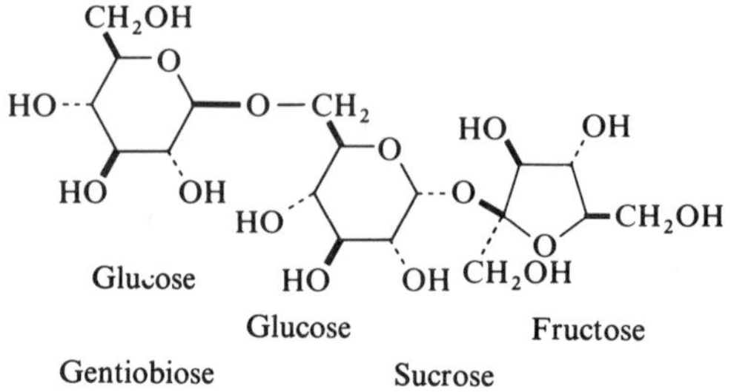

13.11

Stachyose, which occurs in the roots of several plant species, yields sucrose and raffinose on enzymic hydrolysis. Methylation and hydrolysis gives 2,3,4,6-tetra-*O*-methyl-D-galactose, 2,3,4-tri-*O*-methyl-D-galactose, 2,3,4-tri-*O*-methyl-D-glucose, and 1,2,4,6-tetra-*O*-methyl-D-fructose. This means that stachyose is *O*-α-D-galactopyranosyl-(1–6)-*O*-α-D-galactopyranosyl-(1–6)-*O*-α-D-glucopyranosyl-(1–2)-β-D-fructofuranoside (or α-D-Gal*p* 1–6 α-D-Gal*p* 1–6 α-D-G*p* 1–2 β-D-Fru*f*) **13.12**.

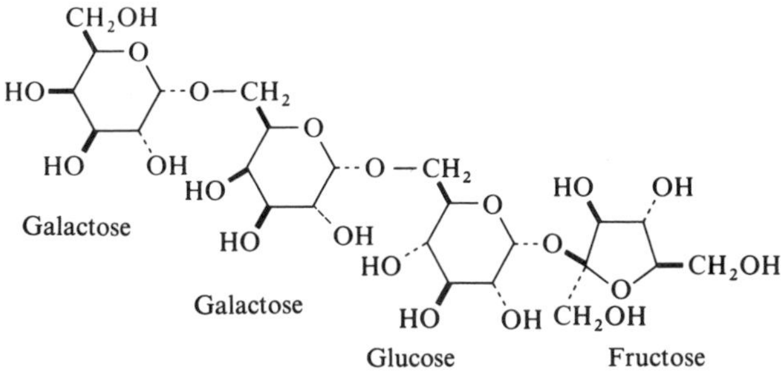

13.12

Schardinger dextrins

These compounds are a group of crystalline homologous oligosaccharides obtained from starch by the action of *Bacillus macerans* amylase and first obtained by Schardinger (1903). Fractionation of the crude product yields the α- and β-dextrins as major products. Application of the usual structural methods showed that the α- and β-dextrins are cyclic molecules of six and seven D-glucose units, respectively, linked α1–4; they are also known as cyclohexa-amylose and cyclohepta-amylose. Because these rather novel molecules lack an end group they are more resistant to acidic and enzymic hydrolysis than their linear counterparts. They are also of interest in that they form complexes with many inorganic compounds, including iodine and salts.

14 Polysaccharides

POLYSACCHARIDES, also called *glycans*, are polymeric substances, the building blocks of which are monosaccharides, joined by glycosidic linkages. They have widespread distribution in nature, in a variety of roles. However, less is known about polysaccharides than about the other two major classes of biological polymers, the nucleic acids and the proteins. This situation is rapidly being remedied.

Cellulose and chitin are structural materials in plants and in some animals, respectively. The main energy reserve materials, starch and glycogen, are polysaccharides. Other more complex polysaccharides are involved in the bacterial cell wall, connective tissue, blood group substances, natural lubricants, and gums.

The monosaccharides that occur as the monomer units vary widely in type. Commonest are the aldoses and ketoses, notably D-glucose and D-fructose, 2-amino-2-deoxy-aldoses, and alduronic acids.

Polysaccharides built from one monomer only are called homo-polysaccharides (or homoglycans); if they are co-polymers they are known as heteropolysaccharides (or heteroglycans).

Some polysaccharides have a linear structure, whereas others are branched, having a main chain, to which smaller chains are linked glycosidically. All types occur naturally; from simple branching (herring-bone type) to highly-branched, bush-like structures.

General methods of structural study

Before being studied, a polysaccharide, like any other natural product, must be isolated from its natural environment. This must be done with the minimum of degradation for otherwise the substance isolated may not be the naturally-occurring one, or at the very minimum may have a much lower molecular weight than in the natural state. The homogeniety of the material must be established.

For a complete structure to be assigned to a polysaccharide, many questions must be answered. The main ones are: how many different constituent monosaccharides does it contain; if more than one, what is the proportion of each and do they occur in regular or random sequence; are the rings pyranose, furanose, or a mixture of the two; are the glycosidic linkages α, β or a mixture; is the overall structure linear or branched; what is the molecular weight?

Some general methods for seeking the answers to such questions will be discussed before dealing with individual substances.

Acid hydrolysis

Hydrolysis of a polysaccharide with aqueous acid will normally break it down to its monomer units, that is, to monosaccharides. These can be identified by paper chromatography or more normally by gas–liquid chromatography after reduction to the corresponding alditol and conversion either to the per-acetate or to the per-trimethylsilyl ether. Careful mild hydrolysis sometimes yields di-, tri-, or even higher oligosaccharides, which can give useful information about the type of linkage between monomer units and also of their ring-size. It should be noted, however, that monosaccharides can combine together under acidic conditions ('reversion') (see p. 23), and so control experiments must always be carried out.

Acetolysis

This refers to treatment of the polysaccharide with acetic acid–acetic anhydride mixtures containing 2–4 % concentrated sulphuric acid. The 1–6 link is the most resistant to acid hydrolysis, but the least stable to acetolysis. Use of both reactions can therefore lead to different fragments from the same polysaccharide.

Methylation

The principle here is similar to that described for the structural study of oligosaccharides (see p. 78). The difficulty is to ensure methylation is *complete*, otherwise the results will be meaningless.

Smith Degradation (1952)

This uses the sequence: periodate oxidation, borohydride reduction, hydrolysis by very dilute aqueous acid. Generally C_2, C_3, or C_4 fragments are obtained, which are characteristic of the types of glycosidic linkage within the polysaccharide. Examples are shown in Scheme 14.1. The acid hydrolysis conditions used are very mild, so that normal glycosidic linkages are generally not broken. This

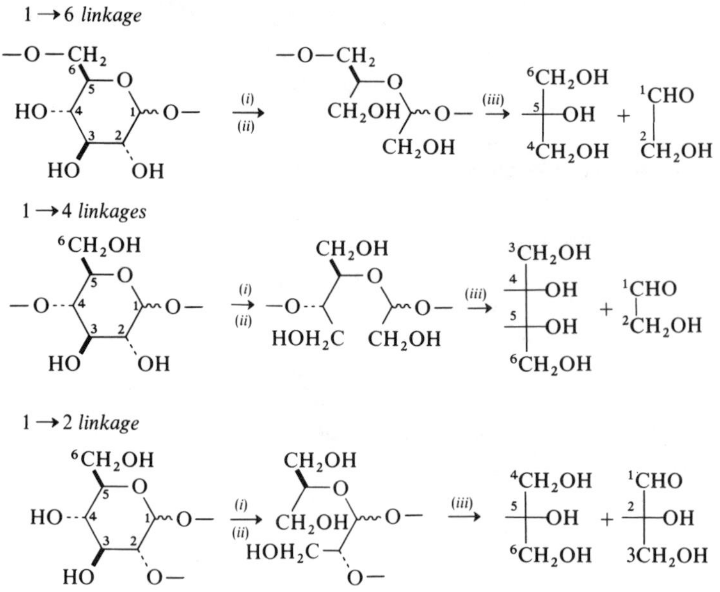

1 → 6 linkage

1 → 4 linkages

1 → 2 linkage

Reagents: (*i*) $IO_4^{\ominus}$; (*ii*) $^{\ominus}BH_4$; (*iii*) very dilute aqueous acid

SCHEME 14.1. Smith degradation of various linkages in pyranose systems; the use of D-glucose units is purely illustrative.

can be useful in identifying 3-linked units, (or 3-blocked units generally), which are not oxidized by periodate (Scheme 14.2).

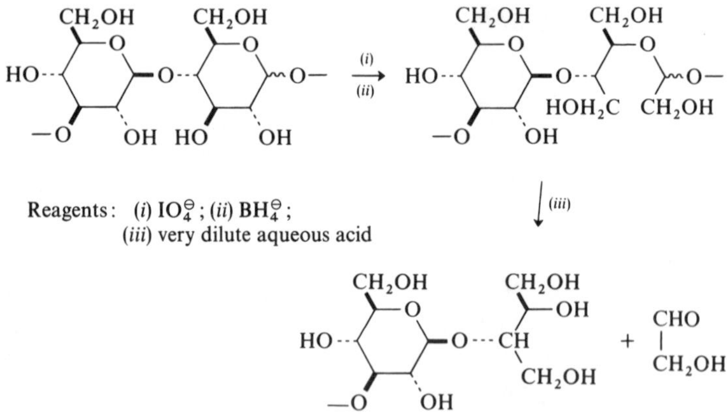

Reagents: (*i*) $IO_4^{\ominus}$; (*ii*) $BH_4^{\ominus}$;
(*iii*) very dilute aqueous acid

SCHEME 14.2. Degradation of a 1 → 3 linked system; the use of D-glucose is again purely illustrative.

Molecular weight

Physical methods, such as osmometry and light-scattering measurements, are much more reliable than most chemical ones; some of the latter will be described later. It should be noted that different physical methods measure different things. For example, osmometry measures the number average molecular weight, whereas light-scattering measures weight average molecular weight.

Plant polysaccharides

Cellulose

Cellulose forms the main constituent of the cell walls of plants. The seed hairs of the cotton plant are almost pure cellulose. The larger fibres are used for spinning into cloth, the shorter hairs (linters) are used for chemical purposes. Other rich sources of cellulose are wood (about 50%), bast fibres, such as flax (80%) and jute (65%), leaf fibres, such as hemp (80%), and cereal straw (45%).

Application of the various methods described above showed that cellulose was a β 1–4 glucan. The anomeric linkage was established by acetolysis which gave cellobiose octa-acetate **14.1** in 50 per cent yield. Very careful acid hydrolysis, followed by chromatogaphy gave the β 1–4 linked oligosaccharides, cellobiose, cellotriose, etc., up to celloheptaose (Zechmeister, 1931, Miller, 1960). There are few solvents for cellulose, one being aqueous cuprammonium hydroxide ('Cupra') with which it forms a copper complex. In a solution of 20% aqueous sodium hydroxide, cellulose swells, and after washing free of alkali, is found to have increased tensile strength, and increased dyeing capacity: it is said to have been 'mercerized' (Mercer, 1850).

14.1   **14.2**

14.3

Cellulose thus has structure **14.2**, or more precisely structure **14.3**, where n is very large. The use of physical methods shows that for native cellulose, n is in the range 2000 to 3000. Chemical methods give much lower values. Most of such methods are based on 'end group assay'; for example methylation.

Complete methylation of cellulose, followed by acid hydrolysis to partially methylated glucoses (Haworth, 1932) yields (Scheme **14.3**)

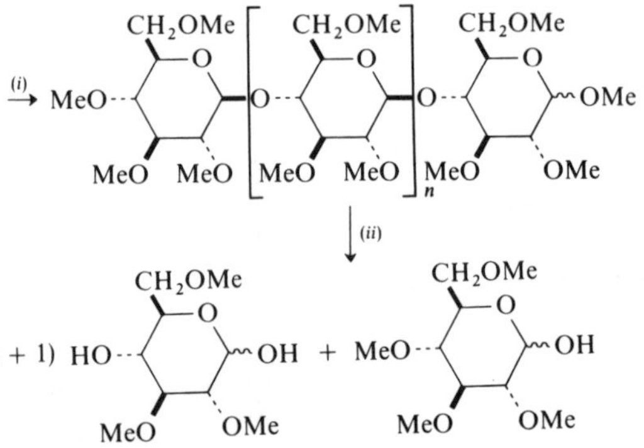

Reagents: (*i*) Me_2SO_4, aq. NaOH; (*ii*) $H_2O^\oplus$

SCHEME 14.3

2,3,6-tri-*O*-methyl-D-glucose and 2,3,4,6-tetra-*O*-methyl-D-glucose in the ratio $(n + 1):1$, hence n. The values obtained were in the range 100 to 200 (*cf.* above). This is because it is not possible to methylate cellulose completely without some degradation. Values for molecular weights based on this sort of method must therefore be taken only as minimum values.

It should be noted that, as with other polysaccharides, cellulose from different sources may have different molecular weights. Indeed samples from the same plant may show considerable variation at different seasons.

X-Ray studies of cellulose have shown that the length of the unit cell is that of two glucopyranose units (Sponsler and Dore, 1926; Meyer and Misch, 1937), showing that in the fibre the chains are aligned along the main axis, as expected. The distance between chains is such that strong inter-chain hydrogen-bonding is possible. Studies on mercerized cellulose show that the length of the unit cell is the same, but that the chains are further apart in accord with the greater accessibility to reagents.

Cellulose samples whose fibres are too short for spinning (for example wood cellulose) can be converted into commercially useful thread, known as *regenerated cellulose*. One process uses precipita-

$$\text{R·OH} \stackrel{(i)}{\rightleftharpoons} \text{RO}^\ominus \stackrel{(ii)}{\rightarrow} \underset{\underset{\text{S}}{\overset{\|}{}}}{\text{R·O·C·SNa}} \stackrel{(iii)}{\rightarrow} \text{R·OH} + \text{CS}_2 + \text{Na}^\oplus$$

R = cellulose

Reagents: (*i*) aq. NaOH; (*ii*) CS_2; (*iii*) $H_3O^\oplus$

SCHEME 14.4

tion from a solution in cuprammonium hydroxide. Another process, which gives *viscose rayon*, uses treatment with sodium hydroxide to give the alkoxide. This reacts with carbon disulphide to give a xanthate salt, a solution of which is forced through a fine orifice into an acid bath, precipitating the cellulose in fibre form (Scheme 14.4). Regeneration through a fine slit gives a thin film, known as 'Cellophane'.

Several cellulose derivatives find commercial use. The triacetate (Tricel or Arnel) has wide textile usage, as does the so-called 'secondary acetate', which has about 2·3 acetate groups per glucose unit; it is also used in the plastics industry. Cellulose trinitrate, 'nitrocellulose', or gun cotton, finds use in explosives. In the textile industry there is much interest in reagents that cross-link cellulose chains, such as formaldehyde or 1,3-dichloropropanol, to confer special properties to the fabric, such as crease resistance.

Starch

Starch is the carbohydrate reserve material of most plants. Its main sources are tubers, such as the potato; and cereals, such as wheat, maize, and rice. It is the main carbohydrate source for animals, including humans. Starch is much more easily hydrolysed than is cellulose and so great care must be taken during isolation and purification to avoid degradation.

Structural studies show that starch, like cellulose, is a glucan, but differs in that it is generally a mixture of two substances *amylose* and *amylopectin*. Several methods are available for fractionation, mostly based on selective precipitation. Starches usually contain about 25% amylose, though selective breeding has led to high amylose starches (about 70%) and also to waxy maize starch, which is almost 100% amylopectin.

Amylose. Chemical and physical methods show that amylose is a linear glucan, with D-glucopyranose units linked α1–4, and with a chain length of 1000–4000 monomer units **14.4**. The amylose

14.4

14.5

backbone can form a variety of helical structures, one of which has about six monomer units per turn **14.5**, and forms in the presence of small molecules which can occupy the space in the centre. The well-known blue starch–iodine indicator colour is due solely to the amylose component. It can therefore be used to determine the amylose content of any starch.

When starch solutions are left to stand for some time, partial precipitation occurs. This is due to the separation of the amylose component and is called *retrogradation*. The helical molecules align themselves by hydrogen-bonding. and when the aggregates exceed colloidal dimensions they precipitate.

Amylopectin. Structural studies show that amylopectin is a branched glucan, with a molecular weight that sometimes exceeds one million. This branched structure is in agreement with the inability to form

films or threads with amylopectin, and also with its non-retro-
gradation. Methylation studies (Hirst, Freudenberg, 1940) showed
the presence of some D-glucopyranose units linked through C-1,
C-4, and C-6. This was confirmed by the isolation of isomaltose **14.6**
and panose **14.7** after partial acid hydrolysis of amylopectin, which

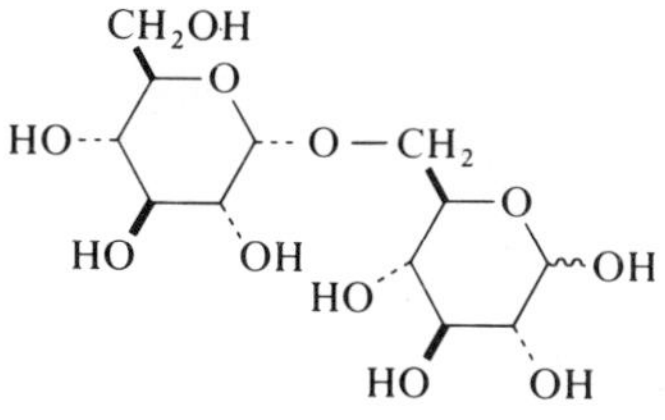

14.6

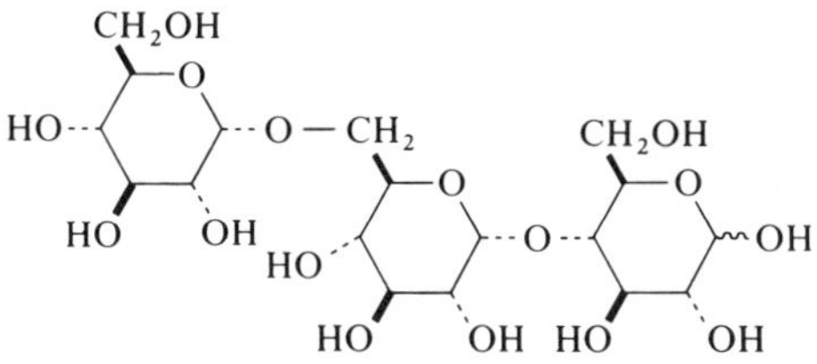

14.7

can therefore be represented as **14.8**. Enzymic studies show that the
branching is completely random (Peat, 1954, Hirst, 1954). The
enzyme β-amylase breaks α 1–4 links but not the α 1–6 ones. Action
on amylopectin therefore degrades the branches to short stubs to
give a material called a *β-limit dextrin* **14.9**.

Inulin

Inulin replaces starch as the carbohydrate reserve material in the
roots of the compositae. Good sources are dandelion, dahlia, and
Jerusalem artichoke tubers. Structural studies show that it is a
fructan linked β2–1. Careful work has shown the presence of α-D-
glucose in the terminal position of the inulin chain **14.10**.

Hemicelluloses

The hemicelluloses are found in association with cellulose in
plant cell walls. The commonest members of this group are the
xylans which are basically β1–4 linked xylopyranose units, but some

14.8

14.9

14.10

also contain other residues such as L-arabinose, D-glucuronic acid and its 4-*O*-methyl ether as part of branch groups.

Other classes are the *mannans*, composed of β 1–4 linked mannopyranose units, found in vegetable ivory, and the *glucomannans*, which occur widely in coniferous woods and are made up of both mannopyranose and glucopyranose units. Both 4-*O*-(β-D-mannopyranosyl)-D-glucose and 4-*O*-(β-D-glucopyranosyl)-D-mannose have been isolated as hydrolysis products.

Gums

Gums are complex polysaccharides exuded by plants. Examples are gum arabic, gum tragacanth, and mesquite gum.

Acid hydrolysis of a gum yields a mixture of monomers: generally D-glucuronic acid, with D-galactose, L-arabinose, D-xylose, and sometimes other sugars. These polysaccharides have highly branched structures.

Polyuronides

The basic building blocks here are uronic acids. One of the better known members of this class is alginic acid which on hydrolysis gives D-mannuronic and L-guluronic acids, the former preponderant. The units are linked 1–4. Sodium alginate is widely used as a thickening agent in the food industry, but it has no nutritional value.

Pectin which occurs in fruit sap, and is the substance responsible for the setting of jam, is a polygalacturonic acid methyl ester in combination with an arabinan and a galactan.

Bacterial polysaccharides

Bacterial cell wall

The polysaccharides of the bacterial cell wall are extremely complex and are usually covalently bound to polypeptide chains. Gram positive bacteria contain *techoic acids* which are polymers of ribitol (or glycerol) phosphate to which are attached D-alanine units **14.11**. Gram negative bacterial do not contain techoic acids in their cell wall.

$$CH_2-O-\overset{\overset{\textstyle O}{\|}}{P}-O-$$

$$-OCOCHCH_3$$
$$-OH \quad NH_2$$
$$-OR$$
$$-CH_2$$

R = H or a sugar

14.11

Another component is a regular repeating co-polymer of 2-acetamido-2-deoxy-D-glucose and *N*-acetylmuramic acid **14.12** linked β1–4. These linear polymers are cross-linked to polypeptide chains by short peptides to give a two dimensional network that provides the strength of the cell wall.

CH$_2$OH   CH$_2$OH

$$-O- \cdots \quad -O- \cdots \quad -O-$$

HO   NHAc O   NHAc
CH$_3$CHCOOH

14.12

Dextrans

These are polysaccharides elaborated by bacterial of the *Leuconostoc* group, and are α 1–6 linked glucans. They have important use as blood plasma substitutes.

Animal polysaccharides

Glycogen

This is the common carbohydrate reserve substance of animals. It is a glucan of similar structure to amylopectin, except that the degree of branching is much greater.

Chitin

This is the polysaccharide component of the shells of anthropods, such as lobsters, crabs, etc. It is readily isolated by dissolving out the calcium carbonate from the shells with dilute acid. Hydrolysis of chitin with hot aqueous hydrochloric acid yields 2-amino-2-deoxy-D-glucose. This, and other evidence, shows that chitin has the structure **14.13**, very similar to cellulose.

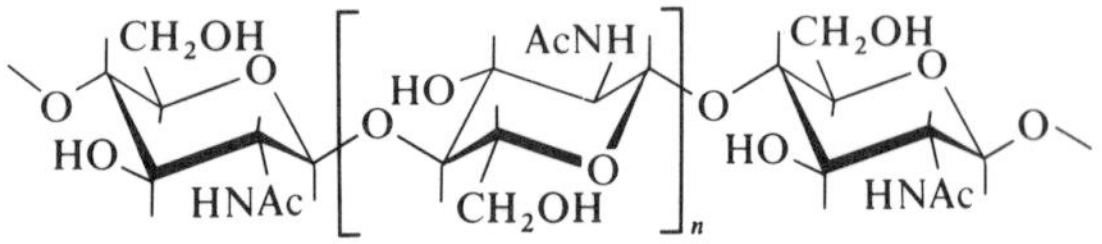

14.13

Sulphated polysaccharides

Many polysaccharides from animal sources have a 2-amino-2-deoxy-aldose as one of the monomers (generally *gluco* or *galacto*) and many of them are sulphated. Examples of such polysaccharides are *heparin* which has the repeating unit **14.14**, (Perlin, 1971), and hyaluronic acid **14.15**.

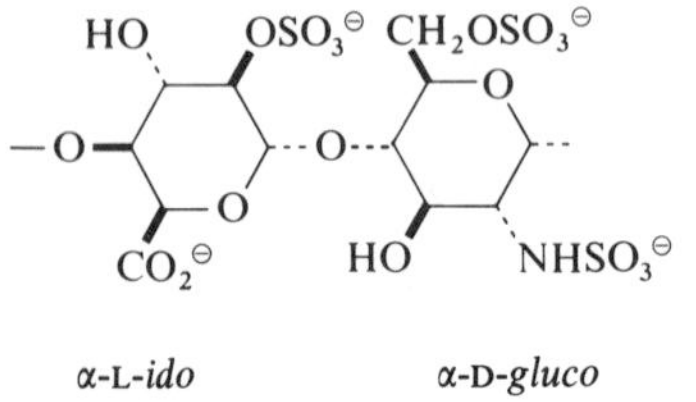

α-L-*ido*   α-D-*gluco*

14.14

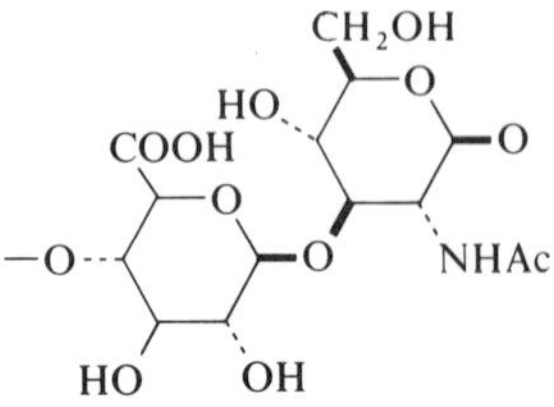

14.15

Polysaccharide gels

Some polysaccharides possess remarkable gelling properties. For example a 0.1 % aqueous solution of agar forms a rigid gel. A satisfactory theory of such behaviour has recently been put forward by Rees (1969). It was shown that polysaccharides such as agars can have double helical structures. Thus it is possible for a network to develop, much like a string bag as shown in **14.16** in which the

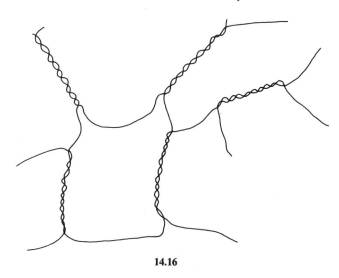

14.16

polymer chains have lengths of ordered helix between unordered sections. Any one chain will link several junction points.

15 Spectroscopic techniques and other physical methods

IN this section it is assumed that the reader has some understanding of the basic principles behind each technique and *only* aspects relevant to carbohydrates will be discussed. Readers without the necessary background should consult a text book such as 'Applications of Absorption Spectroscopy of Organic Compounds', by John R. Dyer.

Nuclear magnetic resonance spectroscopy (n.m.r.)

Proton magnetic resonance spectroscopy (p.m.r.) is perhaps the most useful tool for the structural study of carbohydrate systems at present, though ^{13}C n.m.r. (see below) is being increasingly used.

Figure 15.1 shows the approximate positions of signals for groups commonly encountered as part of carbohydrate systems. The most

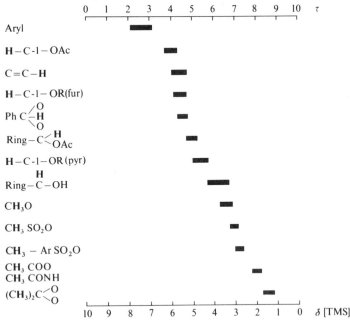

FIG. 15.1. Approximate signal positions in ^1H n.m.r. spectra.

easily assigned signal in the spectrum of an aldose or its derivative is usually the anomeric proton, which, because it is part of an acetal grouping, occurs 0·5–1·00 p.p.m. downfield of the other aliphatic protons. Particular hydrogens on carbons bearing a hydroxyl group can usually be identified by acetylation which moves the signal for the proton downfield by about 1.0 p.p.m.

The most important information that can be obtained from the p.m.r. spectrum is a knowledge of the stereochemistry of molecules, particularly pyranoses, using the spin-coupling values (Js). Indeed the first study of the relationship of J values with stereochemistry in organic molecules was made on monosaccharide acetates (Lemieux, 1958). Vicinal protons in an *ax,ax* relationship show large couplings of 8–10 Hz, whilst *eq,ax* and *eq,eq* arrangements have Js of 2·5–4·5 Hz. Such values are very useful in determining the conformation of a particular monosaccharide derivative.

Consider for example 2,3,4-tri-*O*-acetyl-β-D-xylopyranosyl fluoride **15.1** which could exist in one of the two chair forms **15.2** or **15.3** or less likely in some boat form. The coupling constants for $J_{1,2}, J_{2,3}, J_{3,4}, J_{4,5a}, J_{4,5e}$ are all consistent with **15.3** and are not large enough for **15.2** in which all the vicinal protons are *ax,ax* related, and for which values of $J_{2,3}, J_{3,4}$ etc. can be obtained from the xylopyranosyl per-acetates (see also p. 17).

15.1   **15.2**   **15.3**

Another powerful demonstration of the use of p.m.r. was the demonstration of a conformational equilibrium between the chair forms of β-D-ribopyranose tetra-acetate **15.4** in acetone (Horton, 1969). Variable temperature studies showed that at room temperature the spin-coupling data were intermediate between those expected for **15.5** and for **15.6**; that is a conformational equilibrium

15.4   **15.5**   **15.6**

was involved. At $-60°$ the p.m.r. spectrum was resolved to the sum of the expected spectra for **15.5** and **15.6**: for example, two anomeric signals were observed, one with a large (ax,ax) J value and one with a much smaller J value (eq,ax). At $-60°$ the equilibrium was approximately **15.5**: **15.6** = 1:2. At room temperature the ratio was 55:45.

Examples of the use of p.m.r. in structural studies are given in Chapter 16.

^{13}C magnetic resonance spectroscopy (c.m.r.). At the time of writing (early 1973) there are only a score or so of papers on c.m.r. of carbohydrates compared with the hundreds on p.m.r. of such molecules. It is only in the past few years that natural abundance ^{13}C n.m.r. has become a practical proposition for the organic chemist, following the development of pulsed Fourier transform spectrometers. (For a general account see, for example, Levy and Nelson's *^{13}C N.m.r. for organic chemists*).

One obvious advantage of c.m.r. over p.m.r. is that non-protonated carbon atoms can be investigated, such as carbonyl carbons, quaternary carbons etc. Another is that the ^{13}C resonance range for

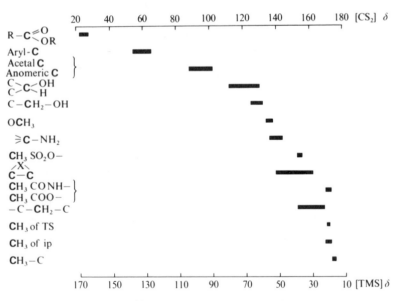

Fig. 15.2. ^{13}C n.m.r. chemical shifts (CDCl$_3$ or D$_2$O).

organic compounds extends over about 200 p.p.m. compared with about 10 p.p.m. for protons. Figure 15.2 gives approximate resonance positions for the type of carbon commonly encountered in carbohydrate systems, but as in p.m.r., other factors such as solvent effects can affect signal positions, so the Figure is meant to serve only as an approximate guide. Note too that the chemical shifts are given with reference to tetramethylsilane, whereas in the literature, carbon disulphide, cyclohexane, and occosionally other solvents are sometimes used as references. A particular carbon bearing a hydroxyl group is best identified by methylation which moves that carbon's signal downfield by 8–10 p.p.m.

One example of the unique use of c.m.r. in monosaccharide chemistry is the study of the solution equilibria of D-fructose (Allerhand, 1971). This, of course, could not be done by p.m.r., as for aldoses (see p. 12), because in a ketose there is no proton on C-2.

Other nuclei. In carbohydrate chemistry the important ones are ^{19}F and ^{31}P. The former has been invaluable in the study of the structure and conformation of fluoro-sugars (Hall, Foster, 1967 on). The latter nucleus can be used to obtain information about phosphate esters.

Infrared spectroscopy

This technique, although much-used, does not present any special advantages to the carbohydrate chemist and it plays the same role as in general organic chemistry. Hydrogen-bonding can be studied using this technique. The ring-size of sugar derivatives can be shown by the presence or absence of certain bands.

Ultraviolet spectroscopy

This region of the spectrum has found little use in carbohydrate chemistry, since neither the free sugars nor their more usual derivatives contain chromophoric groups.

Mass spectrometry

Like infrared spectroscopy this technique is used considerably by carbohydrate chemists, but has no particular advantages to them. Indeed, it generally lacks the ability to differentiate between stereo-isomers and can therefore only yield information about the gross structure of a molecule.

Optical measurements

The optical rotation of a solution of a sugar or a derivative at the sodium D line ($[\alpha]_D^t$ value) provides a physical constant for characterization and an assessment of purity.

Several rules relating rotation to structure have been devised. The most used is Hudson's isorotation rule, which assumes that the molecular rotation ($[\alpha]_D \times MW/100$) of a cyclic sugar is composed of two parts: A, the contribution of C-1, and B, the contribution of the rest of the molecule: it is assumed that changes in one part do not affect the contribution of the other part. The rule does not have general applicability and many anomalies are now known.

Many workers have developed empirical calculations for determining the molecular rotation of free sugars and their simple derivatives (Whiffen, 1956; Bose and Chatterjee, 1958; Lemieux, 1970).

Optical rotatory dispersion and circular dichroism techniques have not been widely applied in carbohydrate chemistry because of the lack of chromophores in the molecules. One useful application is in the determination of the chirality and hence the absolute configuration, of vicinal diols and amino-alcohols by measuring the circular dichroism spectra of their copper complexes (Guthrie, 1968), (see also p. 16).

X-ray spectroscopy

As in other fields of organic chemistry, X-ray spectroscopy has been used to determine structures, conformations and absolute configurations of carbohydrates and their derivatives in the solid state.

16 Structural analysis and synthesis

As in any area of natural product chemistry, substances isolated from their natural source must have their structures rigorously assigned. The principle methods are (1) chemical, particularly devising ways of degrading the molecule into smaller meaningful fragments that can more easily be identified; and (2) physical methods, particularly nuclear magnetic resonance, infrared, and mass spectrometric techniques. The ultimate physical method is, of course, X-ray crystallography. Having deduced a structure it is usual to confirm it by an independent synthesis.

This Chapter selects a few examples of these techniques to illustrate the way in which the major goals of structural analysis and synthesis may be accomplished.

Structural analysis

Garosamine

[D. J. Cooper, M. D. Yudis, R. D. Guthrie, and A. M. Prior, *Journal of the Chemical Society*, (Section C), 1971, 960].

Methanolysis of the chemotherapeutically important gentamicin C complex yields amongst other products the methyl glycoside of an amino-monosaccharide, given the trivial name 'garosamine'. The product was a mixture of α and β anomers (2:1) from the p.m.r. spectrum. Integration of the spectrum showed 17 protons, three of which were exchangeable in deuterium oxide. The mass spectrum showed the molecular ion to be m/e 191 ($C_8H_{17}NO_4$), and a base peak at m/e 160 ($M - \text{OMe}$). Acetylation of methyl α,β-garosaminide with acetic anhydride in methanol gave the N-acetyl derivative (see p. 60) from which the β-anomer was fractionally crystallized and de-N-acetylated with hydrazine hydrate to give pure methyl β-garosaminide. The latter consumed 1·8 mol. of periodate, whereas the N-acetyl compound consumed no oxidant. These results suggest a 3-amino-3-deoxy-pyranose system (see p. 54). The ready methanolysis to give methyl α,β-garosaminide from gentamicin suggests that the amino function is not on C-2 (see p. 60). Study of the p.m.r. spectrum (see Figure 16.1) of methyl β-garosaminide yields partial

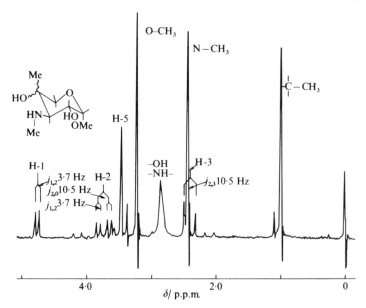

FIG. 16.1. ^{1}H N.m.r. spectrum ($[^2H_6]$benzene; 60 MHz) of methyl 3-deoxy-4-*C*-methyl-3-methylamino-L-arabinopyranoside. (Reproduced with permission from *J. chem. Soc. C*, 1971, 970.)

16.1   **16.2**   **16.3**

structure **16.1** or its enantiomer. The facile, exothermic reaction of methyl β-garosaminide with benzaldehyde to give an oxazolidine **16.2** establishes the *cis* relationship of O–4 and N–3, and hence structure **16.3** or its enantiomer. Consideration of the optical properties of the copper complex of methyl α,β-garosaminide showed conclusively the absolute configuration to be **16.3**.

Everninose

(A. K. Ganguly, O. Z. Sarre, and J. Morton, *Chemical Communications*, 1969, 1488).

Hydrolysis of the antibiotic everninomicin D gave a mixture of products, one of which was everninose ($C_{14}H_{26}O_{10}$), $[\alpha]_D - 74°$

(water). It was a non-reducing sugar, consuming two equivalents of periodic acid, and did not form a trityl derivative. Its p.m.r. spectrum (in pyridine) showed the presence of two anomeric protons ($\tau 4 \cdot 75$ and $4 \cdot 3$) and three OMe groups ($\tau 6 \cdot 65$, $6 \cdot 5$, and $6 \cdot 35$). It also formed a tetra-O-acetyl derivative that had no –OH bands in the i.r. spectrum. Thus the substance is probably a disaccharide linked between the anomeric centres, has four secondary –OH groups (with either three of them adjacent or two pairs of vicinal groups), has no CH_2OH group, and three OMe groups. The mass spectrum of the tetra-TMS derivative showed a molecular ion at m/e 642, in agreement with the above evidence, and also prominent peaks at m/e 335 and 291. The sum of these fragments is $(M - 16)$ and they represent splittings such that either aglycone is lost. Considering all the above data, the partial structure **16.4** can be written.

16.4

Prolonged acid hydrolysis of everninose gave two monosaccharides (M.1) and (M.2), separated by preparative thin layer chromatography. Substance (M.1) was a reducing sugar, had two OMe groups (p.m.r.) and formed a triacetate. The p.m.r. spectrum of the latter compound was identical to the derivative of another natural product curamicose tri-acetate **16.5**. Methylation of the methyl glycoside of (M.1), followed by hydrolysis with aqueous acid gave the known 2,3,4,6-tetra-O-methyl-D-mannose. Thus (M.1) is 2,6-di-O-methyl-D-mannose **16.6**.

16.5   **16.6**

Substance (M.2), a reducing sugar, was converted into a di-n-propyl dithioacetal the mass spectrum of which suggested a 2-O-methyl-pentose. It formed a tri-acetate whose p.m.r. spectrum was completely assignable with the aid of spin-coupling techniques. Key results from the spectrum were the values $J_{2,3}2.9$, $J_{3,4}8.0$, $J_{4,5a}7.0$, and $J_{4,5e}4.5$ Hz which are only compatible with the *lyxo* configuration, that is (M.2) is 2-*O*-methyl-D- or L-lyxose. The absolute configuration was established by the copper complexing method, and by complete methylation followed by acid hydrolysis to give the enantiomer of the known 2,3,4-tri-*O*-methyl-D-lyxose, so that (M.2) is 2-*O*-methyl-L-lyxose **16.7**.

The remaining feature is the mode of linkage of the two moieties **16.6** and **16.7**. This was tentatively assigned by the use of Klyne's rule, and structure **16.8** was proposed for everninose.

16.7   **16.8**

Synthesis

A major problem in the synthesis of all classes of organic compounds is to select blocking groups for protection of some parts of the molecule so that the remainder can be manipulated. With carbohydrates this problem is magnified because the functional groups present are generally all of the same type, namely hydroxyl groups, and so the use of protecting groups and their removal is more sophisticated. Another difficulty in carbohydrate chemistry is that the number of readily available starting materials is small; they are nearly always free sugars such as D-glucose or D-galactose, or the common amino-sugars, like 2-amino-2-deoxy-D-glucose.

Examples of monosaccharide synthesis will be given below. The background to the problem will be briefly stated, the complete synthesis laid out and then comments on the sequences used given.

The synthesis of methyl abequoside

(G. Siewert and O. Westphal, *Annalen*, 1968, **720**, 171).

This simple synthesis is one of many of abequose derivatives in the literature. Abequose is a naturally-occurring deoxy-sugar, namely 3,6-dideoxy-D-*xylo*-hexose, a constituent of polysaccharides of various strains of *Salmonella*.

As in many syntheses we have to start with a derivative of D-glucose, here methyl α-D-glucopyranoside **16.9** (Scheme 16.1). Inspection of the required product **16.11** shows that we have to carry out three changes: (1) convert CH_2OH to CH_3, (2) remove the hydroxyl group at C-3 and (3), invert the hydroxyl group at C-4. The build-up in the synthesis is to **16.10** which possesses the potential for forming an epoxide and has a group at C-6 that is readily converted to CH_3.

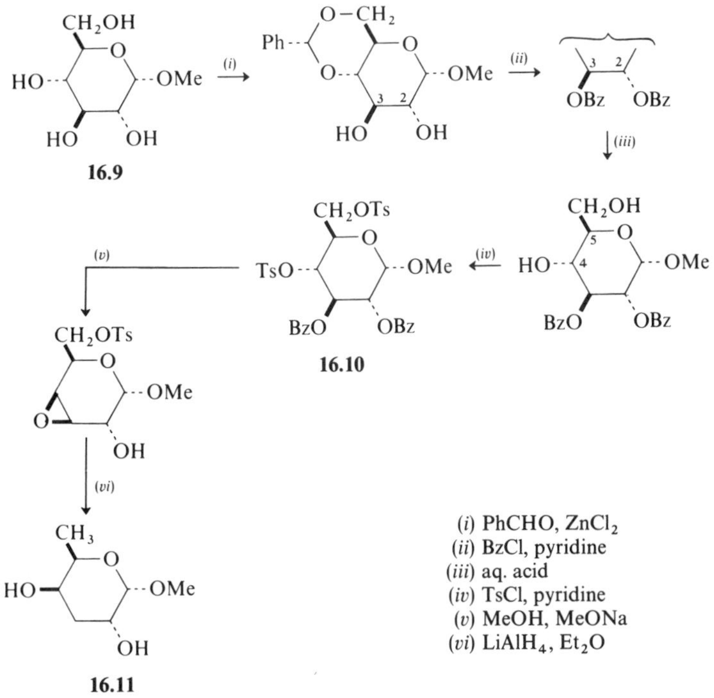

16.9

16.10

16.11

(*i*) PhCHO, $ZnCl_2$
(*ii*) BzCl, pyridine
(*iii*) aq. acid
(*iv*) TsCl, pyridine
(*v*) MeOH, MeONa
(*vi*) $LiAlH_4$, Et_2O

SCHEME 16.1.

Step-by-step analysis: (*i*) This is simple acetal formation to block 4-OH and 6-OH. (*ii*) Protection of 2-OH and 3-OH so that they will not react in step (*iv*). (*iii*) Removal of the acetal blocking group. (*iv*) Tosylation of the now-free 4-OH and 6-OH. (*v*) Treatment of **16.10** with methanolic sodium methoxide first de-benzoylates at O-2 and O-3 leaving at O-3, O-4 a *trans*-hydroxyl-sulphonate system, which under the reaction conditions forms an epoxide by

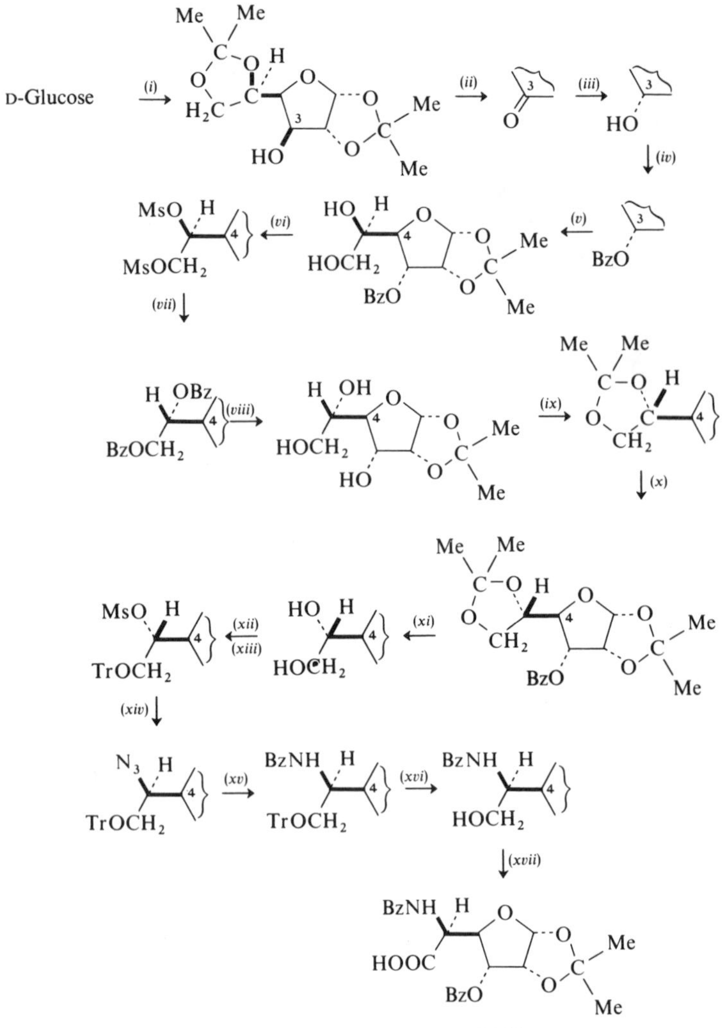

D-Glucose → (i) → [structure] → (ii) → [structure] → (iii) → [structure] ↓ (iv)

(vi) ← [structure] ← (v) ← [structure]

(vii) ↓

[structure] → (viii) → [structure] → (ix) → [structure] ↓ (x)

(xii)/(xiii) ← [structure] ← (xi) ← [structure]

(xiv) ↓

[structure] → (xv) → [structure] → (xvi) → [structure] ↓ (xvii)

[structure]

16.12

(i) Me₂CO, H₂SO₄   (ix) Me₂C(OMe)₂, TsOH
(ii) RuO₄   (x) see (iv)
(iii) NaBH₄   (xi) see (v)
(iv) BzCl, pyridine   (xii) TrCl, pyridine
(v) v. dil. aq. H₂SO₄   (xiii) see (vi)
(vi) MsCl, pyridine   (xiv) NaN₃, DMF
(vii) NaOBz, DMF   (xv) reduction, and then (vi)
(viii) MeOH, MeONa   (xvi) TsOH, Me₂CO
    (xvii) KMnO₄, AcOH, Me₂CO

SCHEME 16.2.

intramolecular attack of C-3-O on C-4-OTs. (*vi*) Lithium aluminium hydride converts the C-6-OTs to C-6-H by hydride ion attack; the same species opens the epoxide stereospecifically to give the required product **16.11**. The free sugar could be obtained by dilute aqueous acid treatment of the methyl glycoside.

Synthesis of the 5-amino-5-deoxy-D-allofuranuronic acid solution

(T. Naka, T. Hashizume, and M. Nishimura, *Tetrahedron Letters*, 1971, 95).

5-Amino-5-deoxy-D-allofuranuronic acid **16.12** is the common sugar moiety of the antifungal nucleoside antibiotics, the polyoxins, used in Japan as agricultural fungicides. This very lengthy synthesis shows the construction of the basic skeleton of **16.12** starting with D-glucose (Scheme 16.2). There are only three key steps: (1) inversion at C-3, (2) the double inversion at C-5 with incorporation of an amino-group, and (3) the oxidation of CH_2OH to COOH. However, seventeen steps are needed to accomplish this so that at the right time *only* the appropriate site in the molecule is available for reaction. *Step-by-step analysis*: (*i*) D-Glucose is protected at O-1, O-2, O-5, and O-6 to leave the 3-OH free for oxidation. (*ii*) Oxidation of 3-OH. (*iii*) Reduction of the carbonyl group, the reagent approaching from the least-hindered β-face of the molecule to give the *allo* configuration. (*iv*) Blocking 3-OH. (*v*) Selective hydrolysis of the 5,6-*O*-isopropylidene group. (*vi*) Mesylation in preparation for S_N2 reaction at O-5. (*vii*) Nucleophilic displacement of the two mesyloxy groups, that at C-5 with inversion to give the L-*talo* configuration. (*viii*) Alkaline hydrolysis of ester groups to give 1,2-*O*-isopropylidene-β-L-talofuranose. (*ix*), (*x*), (*xi*) Steps (*xii*) and (*xiii*) require only 5-OH and 6-OH to be free and since the 3-OH cannot be selectively benzoylated, this sequence has to be used. (*xii*) Selective tritylation of the primary 6-OH. (*xiii*) Mesylation at O-5 in preparation for S_N2 attack. (*xiv*), (*xv*) Nucleophilic displacement of the mesyloxy group by azide ion, reduction and benzoylation; (*xvi*) Detritylation under mild acid conditions. (*xvii*) Oxidation at C-6.

Appendix A
Fischer's proof of the structure
of glucose

EMIL FISCHER's ingenious solution of the problem of determining the structure of glucose and other aldoses and ketoses is illustrated here by his proof of the structure of D-glucose. Modern symbols and names replace those used by Fischer.

Natural glucose was arbitrarily assigned to the D-series and hence it has partial structure **A.1**. (Later work showed that it could be synthesised from D-glyceraldehyde and hence this assumption was fortunately correct.) D-Glucose and naturally-occurring mannose give the same phenylosazone and are therefore C-2-epimers. This enabled structures **A.2** and **A.3** to be proposed for these two sugars, though it is not known which is which. The configurations at C-3 and C-4 have also to be determined.

Treatment of naturally-occurring arabinose with hydrogen cyanide, followed by hydrolysis, gave monobasic acids that were the mirror images of the ones obtained by oxidation of D-mannose and D-glucose, and are thus L-gluconic and L-mannonic acids. Natural arabinose belongs therefore to the L-series. The corresponding D-acids would therefore be obtained from D-arabinose. Also L-arabinose, on oxidation with nitric acid, gave an *optically active* dibasic acid, L-arabinaric acid, showing that the hydroxyl group on C-2 must be on the opposite side of the carbon chain to that on C-5; L-arabinaric acid is therefore **A.4**, and D-arabinose is **A.5**. Since this pentose gives derivatives of both D-glucose and D-mannose, these must be **A.6** or **A.7**.

The two dibasic acids obtained by oxidising D-mannose and D-glucose with nitric acid are **A.8** and **A.9**. Since both acids are optically active neither **A.8** nor **A.9** can be symmetrical, and so the C-4-hydroxyl group is to the right. Therefore D-glucose and D-mannose are **A.10** or **A.11** but it is still necessary to decide which is which.

D-Glucaric acid was also obtained from another aldohexose, gulose, by oxidation with nitric acid. This means that gulose is the same as D-glucose except that the –CHO and –CH$_2$OH groups are interchanged. The two aldoses obtained by doing this to **A.10** and **A.11** are, respectively, **A.12** and **A.13**. But **A.10** and **A.12** are identical and the dibasic acid **A.14** can only be formed by the oxidation of *one* aldohexose, namely **A.10**. The dibasic acid **A.15** can be formed from the two aldohexoses, **A.11** and **A.13**, and is therefore D-glucaric acid. Consequently D-glucose is **A.11**, D-mannose is **A.10**, and L-gulose is **A.13**.

Similar reasoning enabled the structures of all the aldoses to be deduced.

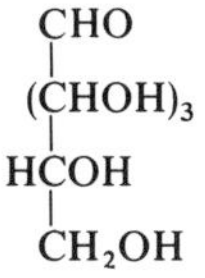

CHO   CHO   CH:N·NHPh   CHO
(CHOH)$_3$ HCOH   C:N·NHPh   HOCH
HCOH   (CHOH)$_2$ → (CHOH)$_2$ ← (CHOH)$_2$
CH$_2$OH   CHOH   HCOH   HCOH
          CH$_2$OH   CH$_2$OH   CH$_2$OH

A.1 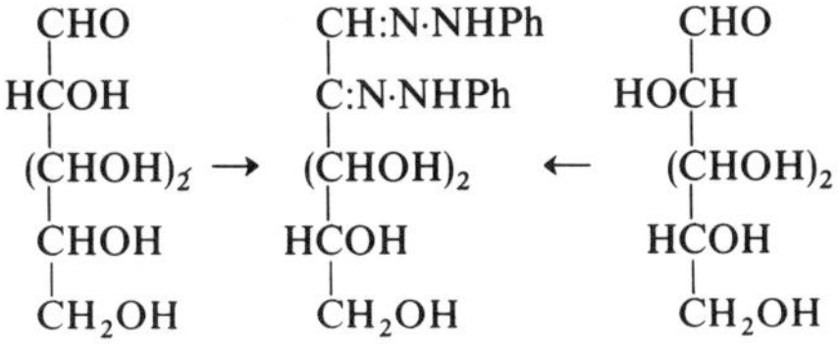   **A.2**   **A.3**

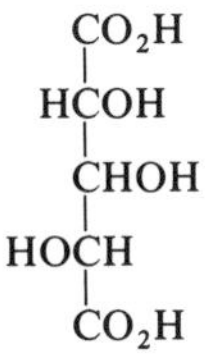

A.4
CO$_2$H
HCOH
CHOH
HOCH
CO$_2$H

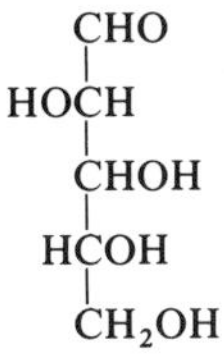

A.5
CHO
HOCH
CHOH
HCOH
CH$_2$OH

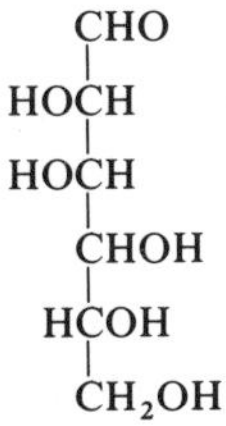

A.6
CHO
HOCH
HOCH
CHOH
HCOH
CH$_2$OH

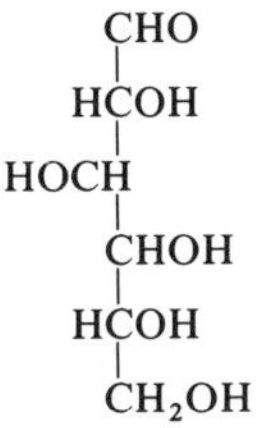

A.7
CHO
HCOH
HOCH
CHOH
HCOH
CH$_2$OH

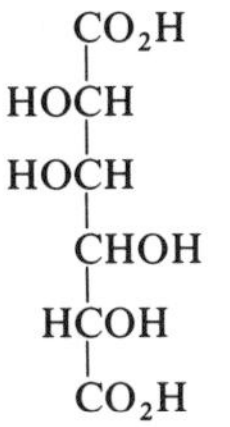

A.8
CO$_2$H
HOCH
HOCH
CHOH
HCOH
CO$_2$H

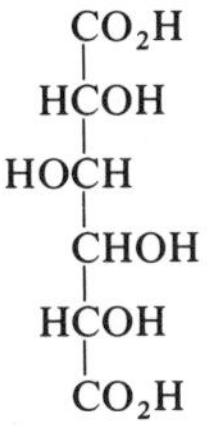

A.9
CO$_2$H
HCOH
HOCH
CHOH
HCOH
CO$_2$H

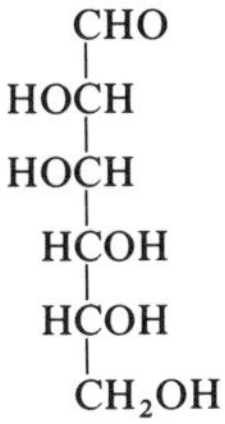

A.10
CHO
HOCH
HOCH
HCOH
HCOH
CH$_2$OH

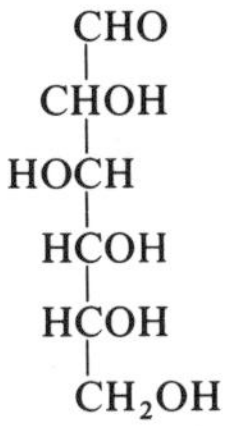

A.11
CHO
CHOH
HOCH
HCOH
HCOH
CH$_2$OH

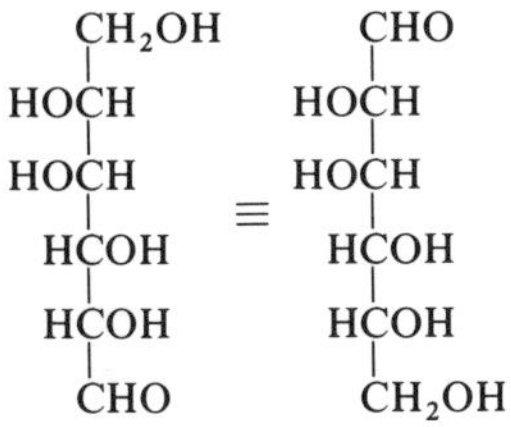

A.12
CH$_2$OH
HOCH
HOCH
HCOH
HCOH
CHO

≡

CHO
HOCH
HOCH
HCOH
HCOH
CH$_2$OH

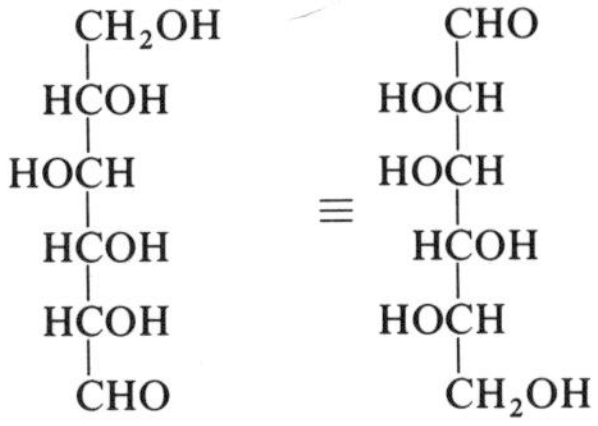

A.13
CH$_2$OH
HCOH
HOCH
HCOH
HCOH
CHO

≡

CHO
HOCH
HOCH
HCOH
HOCH
CH$_2$OH

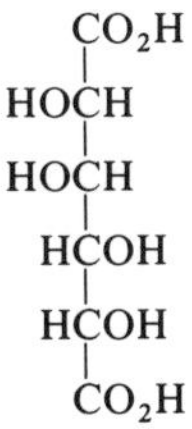

A.14
CO$_2$H
HOCH
HOCH
HCOH
HCOH
CO$_2$H

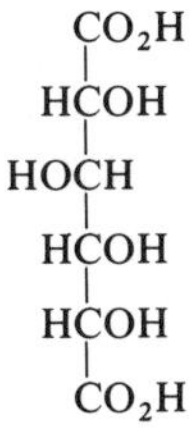

A.15
CO$_2$H
HCOH
HOCH
HCOH
HCOH
CO$_2$H

Appendix B
The carbohydrate literature

So that the interested reader may follow a topic more closely, a short bibliography follows. This is by no means complete, but attention is drawn to the most up-to-date and pertinent reviews and books.

General Texts

'*The carbohydrates*' edited by W. Pigman and D. Horton (Academic Press, 1970–72) in four parts (Vol. IA and B, Vol. IIA and B).

'*Rodd's Chemistry of carbon compounds*', vol. 1F, ed. S. Coffey (Elsevier, 1967). This volume of a large serial work conventionally contains only carbohydrate chemistry.

The monosaccharides by J. Stanek, M. Cerny, J. Kocourek, and J. Pacak (Academic Press, 1963).

The monosaccharides by R. J. Ferrier and P. M. Collins (Penguin Education, 1972); a useful text for graduate students.

Current work

There are many ways of keeping an awareness of the current state of carbohydrate chemistry, through *Chemical Abstracts*, *Current Contents* and so on. Extremely useful publications in the carbohydrate field are the Chemical Society's *Specialist Periodical Reports on carbohydrate chemistry*, which are comprehensive annual surveys. Volume 1 covered the literature for 1967, volume 2, 1968, and so on.

Reviews

Advances in carbohydrate chemistry and biochemistry (formerly *Advances in carbohydrate chemistry*) is the only work devoted solely to carbohydrate reviews: it is published annually by Academic Press. Reviews also appear in the general review journals.

Further Reading

The reviews and texts listed below are not listed chapter by chapter, because some of them are relevant to more than one. They are, however, given approximately in the same sequence as the subject matter of the book. All are in English. '*Advances*' refers to *Advances in Carbohydrate Chemistry and Biochemistry*.

Stereochemistry of carbohydrates by J. F. Stoddart, Wiley-Interscience, 1971.

Mechanism in carbohydrate chemistry by B. Capon, *Chem. Rev.*, 1969, **69**, 407.

Mutorotation of sugars in solution, by H. S. Isbell and W. Pigman, *Advances*, 1968, **23**, 11; 1969, **24**, 14.

Configurational analysis in carbohydrate chemistry, by R. J. Ferrier, *Progress in stereochemistry*, 1969, **4**, 43.

Conformational analysis of sugars and their derivatives, by P. L. Durette and D. Horton, *Advances*, 1971, **26**, 49.

Synthesis of *O*-glycosides, by R. J. Ferrier, *Fortsch. chem. Forsch.*, 1970, **14**, 389.

Cyclic acetals of the aldoses and aldosides, by A. N. de Belder, *Advances*, 1965, **20**, 220.

Cyclic acetals of ketoses, by R. F. Brady Jr., *Advances*, 1971, **26**. 197.

2,5-Anhydrides of sugars and related compounds, by J. Defaye, *Advances*, 1970, **25**, 181.

Oxiran derivatives of aldoses, by N. R. Williams, *Advances*, 1970, **25**, 109.

Alditol anhydrides, by S. Soltzberg, *Advances*, 1970, **25**, 229.

Cyclic acyloxonium ions in carbohydrate chemistry, by H. Paulsen, *Advances*, 1971, **26**, 127.

Sulphonic esters of carbohydrates, by D. H. Ball and F. W. Parrish, *Advances*, 1968, **23**, 233; 1969, **24**, 139.

Some recent neighbouring-group participation and rearrangement reactions of carbohydrates, by J. S. Brimacombe, *Fortsch. chem. Forsch.*, 1970, **14**, 367.

Selected methods of oxidation in carbohydrate chemistry, by S. Hanessian and R. F. Butterworth, *Synthesis*, 1971, 70.

Glycol cleavate and related reactions, by C. A. Bunton, in *Oxidation in organic chemistry*, ed. K. Wiberg, p. 367 (Academic Press, 1965).

The amino-sugars, vol. I and II (each in two parts), ed. R. W. Jeanloz, Academic Press, 1969.

Chemistry of osazones, by H. El Khadem, *Advances*, 1965, **20**, 139.

Chemistry and biochemistry of branched-chain sugars, by H. Grisebach and R. Schmid, *Angew. Chem. internat. Edn*, 1972, **11**, 159.

Unsaturated sugars, by R. J. Ferrier, *Advances*, 1969, **24**, 199.

Nitro-sugars, by H. H. Baer, *Advances*, 1969, **24**, 67.

Comparative chemistry of some amino-glycoside antibiotics, by D. J. Cooper in *Report of IUPAC symposium on antibiotics, Quebec*, 1971, (ed. S. Rakhit).

The synthesis of antibiotic sugars, by J. S. Brimacombe, *Angew. Chem. internat. Edn*, 1971, **10**, 236.

The oligosaccharides, by R. W. Bailey, Pergamon, 1965.

The oligosaccharides, by J. Stanek, M. Cerny, and J. Pacak, Academic Press, 1965.

Sucrose, by D. W. Fewkes, K. J. Parker, and A. J. Vlitos, *Sci. Prog., Lond.*, 1971, **59**, 55.

Sucrose chemicals, by V. Kollonitsch, International Sugar Research Foundation Inc., 1970.

Structure, formation, and mechanism in the formation of polysaccharide gels and networks, by D. A. Rees, *Advances*, 1969, **24**,   255.

Gums and Mucilages, by G. O. Aspinall, *Advances*, 1969, **24**, 333.

Aspects of the structure and metabolism of glycoproteins, by R. D. Marshall and A. Neuberger, *Advances*, 1970, **25**, 407.

NMR spectroscopy in the study of carbohydrates and related compounds, by T. D. Inch, *A. Rev. NMR Spectros.*, 1969, **2**, 35.

Crystal-structure data for simple carbohydrates and their derivatives, by G. Strahs, *Advances*, 1970, **25**, 53.

Mass spectrometry of carbohydrate derivatives, by N. K. Kochetkov and O. S. Chizov, *Advances*, 1966, **21**, 39.

The synthesis of rare sugars, by J. S. Brimacombe, *Angew. Chem. internat. Edn*, 1969, **8**, 401.

Author Index

Subject Index